RENEWABLE MATERIALS AND GREEN TECHNOLOGY PRODUCTS

Environmental and Safety Aspects

RENEWABLE MATERIALS AND GREEN TECHNOLOGY PRODUCTS

Environmental and Safety Aspects

Edited by
Shrikaant Kulkarni, PhD
Ann Rose Abraham, PhD
A. K. Haghi, PhD

First edition published 2021

Apple Academic Press Inc.
1265 Goldenrod Circle, NE,
Palm Bay, FL 32905 USA

4164 Lakeshore Road, Burlington,
ON, L7L 1A4 Canada

CRC Press
6000 Broken Sound Parkway NW,
Suite 300, Boca Raton, FL 33487-2742 USA

4 Park Square, Milton Park,
Abingdon, Oxon OX14 4RN

First issued in paperback 2023

Library and Archives Canada Cataloguing in Publication

Title: Renewable materials and green technology products : environmental and safety aspects / edited by Shrikaant Kulkarni, PhD, Ann Rose Abraham, PhD, A.K. Haghi, PhD.

Names: Kulkarni, Shrikaant, editor. | Abraham, Ann Rose, editor. | Haghi, A. K., editor.

Description: First edition. | Includes bibliographical references and index.

Identifiers: Canadiana (print) 20200370723 | Canadiana (ebook) 20200370979 | ISBN 9781771889278 (hardcover) | ISBN 9781003055471 (PDF)

Subjects: LCSH: Green technology. | LCSH: Nanostructured materials. | LCSH: Nanotechnology—Environmental aspects. | LCSH: Green chemistry.

Classification: LCC TD145 .R46 2021 | DDC 628—dc23

Library of Congress Cataloging-in-Publication Data

Names: Kulkarni, Shrikaant, editor. | Abraham, Ann Rose, editor. | Haghi, A. K., editor.

Title: Renewable materials and green technology products : environmental and safety aspects / edited by Shrikaant Kulkarni, Ann Rose Abraham, A.K. Haghi.

Description: First edition. | Palm Bay, FL : Apple Academic Press, [2021] | Includes bibliographical references and index. | Summary: "Renewable Materials and Green Technology Products: Environmental and Safety Aspects looks at the design, manufacture, and use of efficient, effective, safe, and more environmentally benign chemical products and processes. It includes a broad range of application-based solutions to the development of renewable materials and green technology. The latest trends in the green synthesis and properties of CNs are presented in the first chapter of this book for generating social awareness about sustainable developments. The book goes on to highlight the naissance and progressive trail of microwave-assisted synthesis of metal oxide nanoparticles, for a clean and green technology tool. Chapters discuss green technological alternatives for the global abatement of air pollution, effective use and treatment of water and wastewater, renewable power generation from solar PV cells, carbon-based nanomaterials synthesized using green protocol for sustainable development, green technologies that help to achieve economic development without harming the environment, technical solutions to cut down the quantum of N losses, conventional processing techniques in developing the bionanocomposites as the biocatalyst, and more"-- Provided by publisher.

Identifiers: LCCN 2020048744 (print) | LCCN 2020048745 (ebook) | ISBN 9781771889278 (hardcover) | ISBN 9781003055471 (ebook)

Subjects: MESH: Nanotechnology | Nanostructures | Green Chemistry Technology | Sustainable Development | Renewable Energy

Classification: LCC R857.N34 (print) | LCC R857.N34 (ebook) | NLM QT 36.5 | DDC 610.28/6--dc23

LC record available at https://lccn.loc.gov/2020048744

LC ebook record available at https://lccn.loc.gov/2020048745

ISBN: 978-1-77188-927-8 (hbk)
ISBN: 978-1-77463-777-7 (pbk)
ISBN: 978-1-00305-547-1 (ebk)

DOI: 10.1201/9781003055471

About the Editors

Shrikaant Kulkarni, PhD
Assistant Professor, Vishwakarma Institute of Technology,
Department of Chemical Engineering, Pune, India
E-mail: shrikaant.kulkarni@vit.edu

Shrikaant Kulkarni, PhD, has 37 years of teaching and research experience at both undergraduate and postgraduate levels. He has been teaching subjects such as engineering chemistry, green chemistry, nanotechnology, analytical chemistry, catalysis, chemical engineering materials, industrial organization, and management, to name a few, over the years. He possesses master's degrees in Chemistry, Business Management, Economics, Political Science; an MPhil and PhD in Chemistry; as well as other diplomas in HR, Industrial Psychology, Higher Education, Population Education, etc. He has published over 100 research papers in national and international journals and conferences. He has authored 18 book chapters in CRC, Springer, and Elsevier books. He has edited two books in green engineering and renewable materials, which are in the production stage to be published by Apple Academic Press/CRC Press in 2020–2021. Another three books—on carbon nanotubes for green environment and carbon-based nanomaterials for energy storage and artificial intelligence for chemical sciences—are in the process of development. He has coauthored four textbooks on chemistry as well. His areas of interests are analytical and green and sustainable chemistry. He is a reviewer and editorial board member of many journals in green and analytical chemistry of international repute. He has been invited by UNESCO to give a talk on "Green Chemistry Education for Sustainable Development" at the IUPAC international conference on green chemistry held at Bangkok (Thailand) in September 2018, which was well received. He is an esteemed team member of UNSDG working for the attainment of sustainable development goals. He was appointed as an innovation summit judge in a Conrad challenge competition for teams from across the world, sponsored by NASA. He has been instrumental in formulating and coordinating RIO & COP programs dedicated to sustainable development at his institute by UNCSD. He has worked as a resource person for various national and international events.

Ann Rose Abraham, PhD
Assistant Professor, Department of Basic Sciences,
Amal Jyothi College of Engineering, Kanjirappally, Kerala, India
E-mail: annroseabraham86@gmail.com

Ann Rose Abraham, PhD, is currently an Assistant Professor in the Department of Basic Sciences, Amal Jyothi College of Engineering, Kanjirappally, Kerala, India. Dr. Abraham was involved in teaching and also served as the examiner for valuation of answer scripts of Engineering Physics at APJ Abdul Kalam Kerala Technological University. She has research experience at various national institutes, including Bose Institute, SAHA Institute of Nuclear Physics, UGC-DAE CSR Centre, Kolkata, and has collaborated with various international laboratories, such as the University of Johannesburg, South Africa; Institute of Physics, Belgrade; etc. She is a recipient of a Young Researcher Award in the area of physics. Dr. Abraham has delivered invited lectures and sessions at national and international conferences. During her tenure as a doctoral fellow, she taught and mentored students at the postgraduate level at the International and Inter University Centre for Nanoscience and Nanotechnology, India. She has co-authored many book chapters and co-edited books. She has a number of publications to her credit published in many peer-reviewed journals of international repute. Dr. Abraham has received her MSc, MPhil, and PhD degrees in Physics from the School of Pure and Applied Physics, Mahatma Gandhi University, Kerala, India. Her PhD thesis was on "Development of Hybrid Mutliferroic Materials for Tailored Applications." She has expertise in the field of materials science, nanomagnetic materials, multiferroics, and polymeric nanocomposites.

A. K. Haghi, PhD

*Professor Emeritus of Engineering Sciences, Former Editor-in-Chief,
International Journal of Chemoinformatics and Chemical Engineering
and Polymers Research Journal; Member, Canadian Research and
Development Center of Sciences and Culture
E-mail: AKHaghi@gmail.com*

A. K. Haghi, PhD, is the author and editor of 200 books as well as 1000 published papers in various journals and conference proceedings. Dr. Haghi has received several grants, consulted for a number of major corporations, and is a frequent speaker to national and international audiences. Since 1983, he served as professor at several universities. He is former Editor-in-Chief of the *International Journal of Chemoinformatics and Chemical Engineering* and *Polymers Research Journal;* and is on the editorial boards of many international journals. He is also a member of the Canadian Research and Development Center of Sciences and Cultures (CRDCSC), Montreal, Quebec, Canada. He holds a BSc in urban and environmental engineering from the University of North Carolina (USA), an MSc in mechanical engineering from North Carolina A&T State University (USA), a DEA in applied mechanics, acoustics, and materials from the Université de Technologie de Compiègne (France), and a PhD in engineering sciences from Université de Franche-Comté (France).

Contents

Contributors

Thomas Abraham
School of Chemical Sciences, Mahatma Gandhi University, Kottayam, India

M. A. Asha Rani
Department of Electrical Engineering, National Institute of Technology, Silchar 788010, Assam, India

Bhagyalakshmi Balan
School of Chemical Sciences (SCS), Mahatma Gandhi University, Kottayam 696560, Kerala, India

Pranjit Barman
Department of Chemistry, National Institute of Technology, Silchar 788010, Assam, India

A. Yu. Bondar
M.T. Kalashnikov Izhevsk State Technical University, Izhevsk, Russia

Anu Rose Chacko
School of Chemical Sciences, Mahatma Gandhi University, Kottayam, India

M. Chakkarapani
Department of Electrical and Electronics Engineering, Madanapalle Institute of Technology and Science, Madanapalle 517325, Andhra Pradesh, India

Sijo Francis
Department of Chemistry, St. Joseph's College, Moolamattom, India

V. I. Kodolov
Basic Research, High Educational Centre of Chemical Physics and Mesoscopics, UD of RAS, Izhevsk, Russian Federation
M.T. Kalashnikov Izhevsk State Technical University, Izhevsk, Russia

V. V. Kodolova–Chukhontzeva
Basic Research, High Educational Centre of Chemical Physics and Mesoscopics, UD of RAS, Izhevsk, Russian Federation
Institute of Macromolecular Compounds, Russian Academy of Sciences, Saint Petersburg, Russia

Binila K. Korah
School of Chemical Sciences, Mahatma Gandhi University, Kottayam, India

Ebey P. Koshy
Department of Chemistry, St. Joseph's College, Moolamattom, India

Shrikaant Kulkarni
Department of Chemical Engineering, Vishwakarma Institute of Technology, 666 Upper Indira Nagar, Bibwewadi, Pune 411037, India

Gladiya Mani
School of Chemical Sciences (SCS), Mahatma Gandhi University, Kottayam 696560, Kerala, India

Beena Mathew
School of Chemical Sciences, Mahatma Gandhi University, Kottayam, India

Suresh Mathew
School of Chemical Sciences (SCS), Mahatma Gandhi University, Kottayam 696560, Kerala, India
Advanced Molecular Materials and Research Centre (AMMRC), Mahatma Gandhi University, Kottayam 696560, Kerala, India

R. V. Mustakimov
Basic Research, High Educational Centre of Chemical Physics and Mesoscopics, UD of RAS, Izhevsk, Russian Federation

Yu V. Pershin
Basic Research, High Educational Centre of Chemical Physics and Mesoscopics, UD of RAS, Izhevsk, Russian Federation

Neena John Plathanam
School of Chemical Sciences, Mahatma Gandhi University, Kottayam, India

Mamatha Susan Punnoose
School of Chemical Sciences, Mahatma Gandhi University, Kottayam, India

C. R. Sreerenjini
School of Chemical Sciences (SCS), Mahatma Gandhi University, Kottayam 696560, Kerala, India

I. N. Shabanova
Basic Research, High Educational Centre of Chemical Physics and Mesoscopics, UD of RAS, Izhevsk, Russian Federation

N. S. Terebova
Basic Research, High Educational Centre of Chemical Physics and Mesoscopics, UD of RAS, Izhevsk, Russian Federation

Remya Vijayan
School of Chemical Sciences, Mahatma Gandhi University, Kottayam, India

G. I. Yakovlev
Basic Research, High Educational Centre of Chemical Physics and Mesoscopics, UD of RAS, Izhevsk, Russian Federation
M.T. Kalashnikov Izhevsk State Technical University, Izhevsk, Russia

Abbreviations

AI	*Azadirachta indica*
AM	air mass
AOPs	advanced oxidation processes
AP	ammonium perchlorate
AV	aloe vera
AFM	atomic force microscopy
BDD	boron-doped diamond
CAS	conventional activated sludge
CB	conduction band
CDs	carbon dots
CFRC	coir fiber reinforced concrete
CII	Confederation of Indian Industry
CLIs	crop–livestock integration
CNs	carbon nanomaterials
CNDs	carbon nanodiamonds
CNTs	carbon nanotubes
COD	chemical oxygen demand
CSP	composite solid propellants
CVD	chemical vapor deposition
CCVD	catalytic chemical vapor deposition
DLC	diamond-like carbon
EC	electrochemical
ECM	extracellular matrix
EDS	energy dispersive spectroscopy
EPA	Environmental Protection Agency
ESM	eggshell membrane
FIELD	farm-scale resource interactions, use efficiencies, and long term soil fertility development
FT-IR	Fourier transform infrared
GbE	Ginkgo biloba extract
GFPD	ground fault protection devices
GINC	graphene-iron oxide nano composite
GO	graphene oxide

GTNC	graphene-TiO$_2$ nanocomposite
HIT	heterojunction with intrinsic thin film layer
HOMO	highest occupied molecular orbital
HTT	hydrothermal treatment
IR	infrared
LAL	laser ablation in liquid
LDHs	layered double hydroxides
LIVSIM	LIVestock SIMulator
LUMO	lowest unoccupied molecular orbital
MB	methylene blue
MBR	membrane bioreactor
MFC	microbial fuel cells
MMS	manure management systems
MOF	metal-organic framework
MPECVD	microwave plasma-enhanced chemical vapor deposition
MPPT	maximum power point tracking
MRI	magnetic resonance imaging
MWPCVD	microwave plasma chemical vapor deposition
MWCNTs	multi-walled carbon nanotubes
NCD	nanocrystalline diamond
NIs	nitrification inhibitors
NMR	nuclear magnetic resonance
NTP	nonthermal plasma processing
OCPD	overcurrent protection devices
OFI	*Opuntia ficus-indica*
OSC	organic solar cells
PAHs	polycyclic hydrocarbons
PCB	polychlorinated biphenyls
PECVD	plasma-enhanced chemical vapor deposition
PGNSs	porous graphene-like nanosheets
PL	photoluminescence
PPNDs	photoluminescent polymer nanodots
QY	quantum yield
rGO	reduced graphene oxide
RGs	resource groups
Rh B	rhodamine B
SAED	selected area electron diffraction
SCWG	supercritical water gasification

STED	stimulated emission depletion
SWCNTs	single-walled carbon nanotubes
TEM	transmission electron microscopy
TMOs	transition metal oxides
TP	tea polyphenol
UCPL	up-conversion photoluminescence
UNCD	ultrananocrystalline diamond
UV	ultraviolet
WAO	wet air oxidation
WHO	World Health Organization
WO	wet oxidation
WPO	wet peroxidation
XPS	X-ray photoelectron spectroscopy
XRD	X-ray diffraction
ZC	Zante currants

Preface

"Green products" refers to the promotion of safe, sustainable, and waste-minimizing chemical processes. Green chemistry is a scientific concept that seeks to improve the efficiency with which natural resources are used to meet human needs for chemical products and services. It encompasses the design, manufacture, and use of efficient, effective, safe, and more environmentally benign chemical products and processes. It can ensure eco-efficiency in everything we do, both individually and as a society. Green products also mean protecting and extending employment, expertise, and quality of life.

This book includes a broad range of application-based problems to make the content accessible for professional researchers and postgraduate students. This title expands upon presented concepts with the latest research and applications, providing both the breadth and depth researchers need. The book also introduces the topic of green products with an overview of key concepts. Scientists and graduate students in chemistry will gain a unique insight into the opportunities and challenges facing renewable materials today in its theoretical and practical implementation.

Green nanotechnology is based on the 12 principles of green chemistry to synthesize new nanomaterials to attain, health, economic, environmental, and social benefits. Carbon in its single entity and various forms has been used in technology and human life for many centuries. Since in the prehistoric times, carbon-based materials have been used as writing and drawing materials. In the past two and a half decades, carbon nanomaterials (CNs), especially carbon dots, carbon nanotubes, graphene, fullerenes, and carbon nanodiamond have been used as efficient materials due to their exclusive properties. The new green methods and technologies associated with CNs are put forward by the researchers all over the world have opened exciting opportunities for the revolution of green nanotechnology. Carbon nanomaterials such as graphene, fullerenes, carbon nanotubes, nanodiamonds, and carbon dots synthesized from renewable organic resources have gathered a considerable amount of attention due to their outstanding properties and applications. The latest trends in the green synthesis and properties of the CNs are presented in the first chapter of this book for generating social awareness about the sustainable developments.

The naissance and progressive trail of microwave-assisted synthesis of metal oxide nanoparticles, a clean and green technology tool have been highlighted in Chapter 2. This easy, facile, and fast environment friendly method ensures clean and controllable material production at the minimal expense of time and chemicals. Microwave radiations act as a nonconventional heating source and several metal oxide nanoparticles and hybrid materials have been developed for photocatalysis, propellants, and supercapasitor applications. The adoption of microwave synthesis route is capable of fabricating fine quality products with significantly improved uniform and fine nanostructures with precise control over chemical composition surface area and interfacial characteristics. In metal oxide nanoparticle synthesis, controlling the nanostructuring of the metal oxide nanoparticle is indeed of great importance and microwave-assisted synthesis of metal oxide nanoparticles transpires as a productive and effective tool in line with green technology and environment safety.

Air pollution seems to be the major unsolved problem of the modern era. Chapter 3 deals with different causes of air pollution and proposes some possible remedies. The green technological alternatives for the global abatement of air pollution are also suggested. The health problems associated with air contamination were also discussed.

Wastewater originated from various industries like textile, agriculture, food, petrochemical, polymer, pharmaceutical, etc. contains a large number of contaminants of oil and salt of inorganic and organic compounds. When this wastewater released into the ecosystem without any appropriate treatments causes major ecological issues with high environmental impacts. Also, the natural fresh water resources are getting depleted because of the increased demand for fresh water supply. There are different physical, chemical, and biological methods developed for the treatment of water, but these methods cannot abolish the contaminants. And also most of these conventional methods are very expensive. The development of green technology for water treatments has received enormous interest over recent years due to its significant advantages to the environment, society, and economy. In Chapter 4, we discuss the various green technologies for the treatment of water and wastewater.

Concern over the limited stock of conventional energy sources such as coal and other petroleum products has fuelled efforts toward the development of renewable sources of energy that have a lesser footprint on the environment. Materials and technology play a vital role that can offer promising solutions to achieve renewable and sustainable pathways for the future. Of these renewable sources, solar radiation is the most abundant and freely available one, and can be directly harnessed by the use of Photovoltaic (PV) modules. This chapter on the whole discusses about renewable power generation from solar PV cells. The renewable power generation scenario, working of a solar cell, model of solar PV module, impact of photovoltaic cell material in PV characteristics, the effect of DC link capacitor material in power converters, challenges involved in solar power generation and their mitigation techniques; and green materials using green chemistry for fabricating the future solar PV cell and DC link capacitors are included in Chapter 5.

Carbon-based nanomaterials are the most valuable materials used in the modern field due to its high potential of application in almost all areas of living. Increase in crisis for energy and environmental degradation are the foremost challenges of using nanotechnology for the sustainable development. Here lies the importance of green nanotechnology, which focuses on the use of green natural precursors for the development of eco-friendly processes and products. In this scenario, researchers have started the use of renewable, inexpensive, and abundant carbon-based nanomaterials for the sustainable development. Even though there are enormous applications of carbon-based nanomaterials in various fields, this chapter proceeds with the applications of green synthesized nanomaterials for future perspectives. The major applications of carbon-based nanomaterials outlined involve prevention of environmental degradation, improvement of public health, energy efficiency, optimization, and industrial development. Green synthesized carbon nanomaterials such as carbon dots, carbon nanotubes, graphene, fullerenes, and nanodiamonds are given special attention due to their excellent and efficient properties. Chapter 6 provides the reader the current progress, highlighting the application in environment and energy-related fields of the carbon-based nanomaterials synthesized using green protocol.

In recent times, environmental and climate problems are one of the daunting issues across the world. Economic development without considering the environmental concerns causes the depletion of natural resources particularly water resources and air. This results in extensive ecological, financial, and social impairment on a global level. Therefore, it is very essential to enhance the usage of renewable resources and reduce the usage of nonrenewable resources to protect the environment for future generations. This reiterates the need to develop green technologies, which are required to achieve economic development without harming the environment. In this chapter, we discussed the synthesis of renewable and sustainable materials by different methods and their application in the development of green technology. The synthesis of some renewable and sustainable materials by different methods and their application in the development of green technology is presented in Chapter 7.

Nutrient cycle is made up of two important geochemical fluxes namely, nitrogen and phosphorous. The nutrient cycle asks for timely action on the environmental policy front in terms of soil management, farming systems, the sewage treatment, etc. The nutrient cycle unfortunately is not getting the necessary attention it deserves as the nutrients policy is vulnerable. Nutrients such as nitrogen (N) and phosphorus (P) are essential for the growth and development of organisms. Ecosystems are responsible for controlling the flows and concentration levels of nutrients through a host of complex processes including biodiversity. Nutrients cycles have so far been modified to a substantial extent by human intervention mainly agriculture over the time, with its own consequences for not only a range of ecosystems but also to human well-being. The nutrients absorption and retention capacity of terrestrial ecosystems supplied by way of fertilizers or deposition has been depleted because of many large ecosystems into large scale but low diversity agricultural systems. Further with the reduction in the buffering capacity of ecosystems like riparian forests, wetlands, and estuaries, the nutrients in excess leach into groundwater, rivers, and lakes and are subsequently transported to coastal ecosystems. Although agriculture intensification is the only option to meet the future food demands and to check conversion of land from natural vegetation to agriculture, excessive flows of N have contributed to eutrophication, acidification of freshwater bodies, and coastal marine ecosystems. Nitrogen losses promote global warming and, to a certain extent, are instrumental in creating ground level ozone and depletion of stratospheric ozone layer.

Chapter 8 take as overview of the sources of N losses at the different stages of N cycling in agroecosystems, technical solutions to cut down the quantum of N losses, and throw light on the analysis supported by integrated tools and indicators in two contrasting situations: a low-input and a high-input system from the developing and an industrial world, respectively. Moreover, the research needs for better assessment of the expected benefits to be achieved at a global level, in land productivity and erosion in environmental impact by improving nutrient cycling management discussed too.

Bionanocomposite materials by virtue of their strength in terms of structural diversity hold lot much of promise and potential in terms of widespread applications in diverse fields ranging from sensing to energy production. The structural diversity can be put to advantage ranging from CNT's to collagen. The diversity can further offer host of combinations of biomaterials derived from bionanocomposites. The compositional diversity is further of immense interest in designing materials with requisite shape, size, geometry, morphology to meet the specific challenges in biocatalysis on demand. However, structural diversity may lead to varied expectations often with lexicon and evolution of a lot much of literature. However, these materials are yet to be explored to their full potential. It is attributed to the nanotoxicity regulatory constraints and disparity in performance in terms of specificity in biocatalytic activity. However, biocatalysts with well-defined architectures exposed to chemical environments in tune with their biological activity can help increase yield by way of enhancing the substrate or mediator diffusion. Further, the right kind of architecture may present a soundness in stability in physicochemical conditions which otherwise may stifle the performance of catalysts.

Chapter 9 emphasizes upon conventional processing techniques in developing the bionanocomposites as the biocatalyst. Further, it explains innovative processing technologies like electrospinning or bioprinting in order to shape living matter which could, in our knowledge, hasten the spectrum of applications of bionanocomposite materials for biocatalysis.

In Chapter 10, a broad study on magnetic metal carbon mesocomposites green synthesis peculiarities with point of chemical mesoscopics view is presented.

The role of nanostructures activity in selected green materials modification is investigated in Chapter 11.

CHAPTER 1

Green Synthesized Carbon-Based Nanomaterials: Synthesis and Properties

BINILA K. KORAH, NEENA JOHN PLATHANAM, ANU ROSE CHACKO, MAMATHA SUSAN PUNNOOSE, THOMAS ABRAHAM, and BEENA MATHEW[*]

School of Chemical Sciences, Mahatma Gandhi University, Kottayam, India

[*]*Corresponding author. E-mail: beenamscs@gmail.com*

ABSTRACT

Green nanotechnology is based on the 12 principles of green chemistry to synthesize new nanomaterials to attain, health, economic, environmental, and social benefits. Carbon in its single entity and various forms has been used in technology and human life for many centuries. Since in the prehistoric times, carbon-based materials have been used as writing and drawing materials. In the past two and a half decades, carbon nanomaterials (CNs), especially carbon dots (CDs), carbon nanotubes (CNTs), graphene, fullerenes, and carbon nanodiamond (ND) have been used as efficient materials due to their exclusive properties. The new green methods and technologies associated with CNs are put forward by the researchers all over the world have opened exciting opportunities for the revolution of green nanotechnology. CNs such as graphene, fullerenes, CNTs, NDs, and CDs synthesized from renewable organic resources have gathered a considerable amount of attention due to their outstanding properties and applications. The latest trends in the green synthesis and properties of the CNs are presented here for generating social awareness about the sustainable developments.

1.1 INTRODUCTION

Carbon which is the main component of living organisms has the inherent potential of combining with other atoms forming stable useful compounds and making it distinctive in all aspects. This uniqueness of carbon has led to the creation of numerous stable forms in all dimensions including carbon dots (CDs), nanotubes, graphene, graphene oxide (GO), fullerenes, and nanodiamond (ND). The use of carbon-based materials in all fields started long time before and continues to be so in a faster rate. The high cost of raw material and the pollution affected to nature are the two main obstacles that minimize the growth of carbon-based nanomaterials toward sustainable development. In most cases, we depend on petroleum and fossil fuel based-precursors which are not eco-friendly and also have a high chance of depletion in the near future. Researchers have recognized this and started developing carbon based nanomaterials from natural precursors, which are renewable, inexpensive, and environment friendly.

This chapter deals with the synthesis and properties of carbon-based nanomaterials from green precursors including plants, microbes, biomolecules, and minerals. The use of these sources as the starting material and the synthesis methods adopted offers many advantages over other conventionally used chemical methods. Among the different carbon-based nanomaterials, the green synthesis and properties of CDs, carbon nanotubes (CNTs), graphene, GO, fullerenes, and NDs are given special emphasis in the following sections (Fig. 1.1). We hope that this chapter leads a brief viewpoint on the properties and future advantages of carbon-based nanomaterials synthesized using green routes.

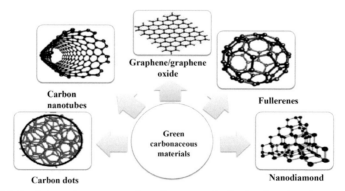

FIGURE 1.1   Schematic representation of green synthesized carbon-based nanomaterials.

1.2 CARBON DOTS (CDs)

The preceding years have witnessed the development of CQDs, C-dots, or CDs outperforming other members of the carbon family. CDs have attracted a lot of attention due to their chemical inertness, resistance, high stability to photo-bleaching, bright fluorescence, low toxicity, excellent aqueous solubility, and biocompatibility. These accountable characteristics have caused them to be applied several fields including sensing,[1] bioimaging,[2] optoelectronic,[3] and medicine.[4] CDs are primarily zero-dimensional nanoparticles, defined by a quasi-spherical morphology and the characteristic size of <10 nm.[5] In 2004, Xu et al. discovered fluorescent carbon nanoparticles inadvertently by an electrophoretic refinement of single-walled carbon nanotubes (SWCNTs).[6] Quantum-sized CDs were reported by Sun et al. in 2006 from the aqueous solution of the PEG_{1500N} with carbon nanoparticles.[5]

In the beginning, the prepared CDs were of low quantum yield (QY) and limited solubility because the synthesis of CDs was limited to carbonaceous material. The recently developed CDs possess enhanced QY and high solubility. The green synthesized CDs involve toxic free precursors which boost up innovation by reducing pollution. The term "green precursors" refers to substances that are either naturally occurring or are derived from renewable natural products or processes.[7] The actual carbon sources of CDs may be from artificial or natural sources. In fact, natural commodities have some advantages over synthetic carbon sources in the preparation of biomass CDs, such as low cost and high abundance. In recent years, many precursors have been identified for synthesis of CDs including graphite,[8] grapheme,[9] C_{60},[10] multi-walled carbon nanotubes (MWCNTs),[11] saccharides,[12] amino acid,[13] citric acid,[14] lauryl gallate,[15] ammonium citrate,[16] ethylene glycol,[17] benzene,[18] phenylene diamine,[19] phytic acid,[20] EDTA,[21] thiourea,[22] human derivatives,[23] animal derivatives,[24] fruit sources,[25] vegetables,[26] leaves,[27] and waste materials.[28]

Herein, we highlight the recent advances in CDs with particular emphasis on the synthetic methods and physiochemical properties. Despite some noteworthy advances, the large-scale utilization of the sustainable resources for fabricating CDs with desirable properties by facile methods is still a great challenge.

1.2.1   SYNTHETIC METHODS

Currently, CDs can be prepared by two potential approaches such as bottom-up and top-down methods (Fig. 1.2). The former strategies incorporate decomposition of small molecules by thermal methods,[29] microwave pyrolysis,[30] chemical treatment,[31] carbohydrates carbonization,[32] electro deposition,[33] and the latter involves, the extraction of carbon nanoparticles from large carbon structures of grapheme,[34] or the laser ablation of graphite or candle soot,[35] electrooxidation,[36] and arc discharge.[5] Among different physicochemical methods detailed up until this point, the hydrothermal method is the most preferred being basic one pot, financial, and green technique for huge-scale synthesis of CDs. These roughly categorized bottom-up and top-down methods are subject to certain changes during preparation including heteroatom doping (nitrogen, boron, and sulfur) and surface modification.[37]

Natural CDs are usually made from food waste, plant waste, and animal waste, which can be synthesized by different approaches. Most studies have attempted to develop an easy, efficient, size controlled, low-cost large-scale preparation approaches for integrating high-quality CDs.[38] Hence, the development of such stable potential CDs via green approaches has become more critical. Here we briefly describe the synthesis of stable CDs via green methods.

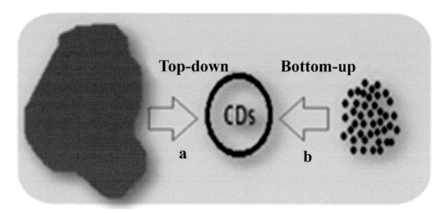

FIGURE 1.2 Conceptual depiction of strategies for the preparation of C-dots. (a) Top-down approach and (b) bottom-up approach.

1.2.1.1 HYDROTHERMAL TREATMENT

Hydrothermal treatment (HTT) or solvent thermal carbonization is a low-cost, environment friendly, and non-harmful route for the manufacture of novel carbon-based materials from different green precursors, and is considered direct and effective. It is widely used because of its easy control of the reaction, high reactivity of reactants, and low energy loss.[39] CDs were prepared via HTT from many precursors such as neem leaves, fenugreek,[40] orange juice,[41] *Jinhua bergamot,*[42] hair,[23] and so on. Usually, the setup is simple and the resulting particle is almost uniform in size with a high QY. In a typical approach, small organic molecules dissolved in water or an organic solvent form the reaction precursor, which was then transferred to a Teflon-lined stainless steel autoclave. The molecules have fused together at relatively high temperatures to form carbon seed cores and then become CDs with a particle size of less than 10 nm.[43]

From the studies so far, four green precursors have a QY of more than 50% through HTT. Showing, Mathew et al. synthesized white photo luminescent CDs with a QY of 63% and size of about 4–5 nm in one step by HTT of areca nut husk extract followed by centrifugation and filtration.[44] Suvarnaphaet et al. prepared blue photoluminescent CDs from limeade via a single-step HTT, whose QY exceeding 50% for the 490 nm applied in detection of Fe^{3+} in bioimaging.[45]. Ren et al. reported CDs with QY of 53% from natural spinach by a hydrothermal method and it can be used for fluorescent ink, biosensors and bioimaging.[46] Yu et al. synthesized water-soluble fluorescent CDs under hydrothermal conditions with a plant of Jinhua bergamot as a carbon source with a QY of 50.78% used for detection of Hg^{2+} and Fe^{3+} with high selectivity, fast response, and low cost.[42] In fact, it is the effortless manufactured procedure and controllable heteroatom doping that makes this technique a promising way to deal with the plan and create novel electrocatalyst with tunable doping composition and electronic structures.

1.2.1.2 MICROWAVE TREATMENT

Among the bottom-up approaches, the method of microwave treatment has been well established due to rapid synthesis and commercialization. The microwave method uses microwave energy to break down the chemical bonds of the substrate by exposing them to electromagnetic radiation in the microwave frequency range.[30] This simple method provides outstanding characteristics such as shorten the reaction time and provide simultaneous homogeneous heating which leads to uniform size distribution of CDs, cost effectiveness, and low energy consumption. Wang et al. designed a "green", rapid, eco-friendly, and waste-reused approach to synthesize fluorescent and water-soluble CDs with QY of 14% from the eggshell membrane (ESM) ashes allowing to a microwave-assisted process used for detection of Cu^{2+}.[47] CDs prepared by microwave-assisted treatment of several precursors such as table sugar,[48] naked oats,[49] flour,[50] sesame seeds,[51] and rose flowers[52] with QY of 2.5, 3, 5.4, 8, and 13.5%, respectively, can be used as a sensor for Pb^{2+}, bioimaging, detection of Hg^{2+}, selective sensing of Fe^{3+}, and detection of tartrazine, respectively.

However, it is difficult to finely control the synthesis process (e.g., irradiation power and absorption properties of precursors) with this technique, which could lead to poor reproducibility and low quality, may have irregular shape, large size distribution, and aggregation since the precursor solution is completely dried eventually. In addition, microwave radiation will also cause certain harm to human beings.[13,17,30]

1.2.1.3 PYROLYSIS

Pyrolysis is the thermal breakdown of materials in an inert atmosphere at high temperatures. External heat can be used to dehydrate and carbonize precursors and convert them into CDs. This method offers the advantages of easy operation, solvent-free approach, wide tolerance of precursors, short response time, low cost, and scalable production.[3,53]

Thakur et al. synthesized CDs via a facile one-step pyrolysis method from *Citrus limetta* waste pulp with the highest QY of 63.3% and great optical and structural quality. Multifunctional aspects of synthesized CDs for photo electrochemical (EC) water splitting, photo catalytic methylene blue degradation, Fe(III) ion sensing, bacterial activity (against *E. coli* and *S. aureus*), and bioimaging with excellent performance.[54] Watermelon peel,[55] lychee seeds,[56] black soya beans,[57] papaya waste pulp,[58] peanut shells,[36] and all of these precursors forms CDs via pyrolysis method with QY of 7.1%, 10.6%, 38.7%, 23.7%, 10.6%, respectively. The obtained CDs possessed good water dispersibility, excellent chemical, and optical stabilities, negligible toxicity, and preferable biocompatibility.

1.2.1.4   EXTRACTION METHOD

In fact, there are CDs in nature. Some researchers isolated CDs in nature through simple centrifugation and purification. This synthesis strategy does not require complex treatment.[38] Cao et al. filtered the Chinese mature vinegar through filter paper to remove large agglomerates from the liquid. Further filtration using membrane and freeze-drying was done to obtain natural CDs with a QY of 5.71%.[59] The existence of CDs in beer was confirmed by Wang et al. They used a rotary evaporator for simple condensation followed by column chromatographic separation. Finally, CDs with 7.39% of QY were obtained by lyophilization.[60] CDs in the daily diet are nontoxic and safe to use as the cytotoxicity study demonstrates. Although the treatment of extraction can avoid the synthesis process, this approach has some limitations. For example, particle sizes of extracted CDs are not uniform.

1.2.1.5   ACID OXIDATION

Acid oxidation treatment is widely used to disperse and decompose bulk carbon to nanoparticles, as well as introduce hydrophilic groups, for example, a hydroxyl group or carboxyl group on its surface to obtain CDs that significantly improve the solubility and fluorescence properties.[43] Myint et al. reported the synthesis of water-soluble and environment friendly fluorescent CDs using the renewable resource lignin as a precursor, which is a waste emitted from pulp and paper industries by acid oxidation (H_2SO_4:HNO_3 = 1:3, v/v) treatment. Moreover, the CDs of 13% QY exhibited excellent stability and cellular imaging capability with very low cytotoxicity.[61] Bhamore et al. instigated water-dispersible multicolor emissive CDs that were obtained from Manilkara zapota fruits and it well-tuned by acid oxidation, which results to generate blue, green, and yellow CDs with 5.7%, 7.9%, and 5.2% of QY. Because of ultra-small size and biocompatibility, three C-dots act as promising bioimaging agents for imaging of *E. coli, Aspergillus aculeatus,* and *Fomitopsis sp* cells.[62]

However, there are few reports on the acid oxidation of bulk materials to CDs.

1.2.1.6   OTHER SYNTHESIS METHODS

Except for the above regular approaches, some unique methods were also used to synthesize CDs with satisfactory results. For example, Wang et al. prepared amphiphilic CDs through plasma treatment with a yield of 5.96%. As carbon precursors of CDs, chicken eggs are rich in nitrogen and have the potential to allow self-passivation and doping.[63] Hsu et al. synthesized highly water-soluble, biocompatible, and photo-luminescent CDs having an average diameter 3.4 ± 0.8 nm and QY of 4.3% are obtained from used green tea through grinding, calcination, and centrifugation.[64] Meral et al. proposed liquid nitrogen-assisted synthesis of CDs from *Blueberry* and their performance in Fe^{3+} detection.[65]

Above mentioned synthetic methods are off under the bottom-up strategy because bottom-up methods can be utilized to synthesize CDs from special organic molecules. All utilize various molecular precursors to controllably synthesize CDs with functional groups within one step. Table 1.1 shows the literature on organic precursors used in the bottom-up synthesis of organic CDs, synthetic conditions, and the QY studied in each research.

TABLE 1.1   Synthetic Methods of Organic CDs and Quantum Yields.

Synthetic method	Precursors	Procedures	QY (%)	Ref.
Hydrothermal	Hair	120°C, 24 h, dialysis	29.0	[23]
	Apple juice	150°C, 12 h, filtration, washed with CH_2Cl_2, centrifugation, dialysis	4.3	[25]
	Daucus carota Subsp. sativus Roots	170°C, 12 h, centrifugation, dialysis	7.6	[26]
	Aloe	180°C, 11 h, filtration	10.4	[27]
	Neem leaves	80°C, 1 h, filtration, centrifugation, 300°C, 8 h, centrifugation	41.2	[40]

TABLE 1.1 *(Continued)*

Synthetic method	Precursors	Procedures	QY (%)	Ref.
	Fenugreek	80°C, 1 h, filtration, centrifugation, 300°C, 8 h, centrifugation	38.9	[40]
	Orange juice	120°C, 2.5 h, washed with CH_2Cl_2, centrifugation	26.0	[41]
	Jinhua bergamot	200°C, 5 h	50.6	[42]
	Areca nut husk Extract	120°C, 3 h, centrifugation, filtration	63.0	[44]
	Limeade	106°C, 1 h, filtration, sonication 1 h, filtration, annealed at 200°C, 2 h, sonication, filtration.	53.6	[45]
	Spinach	189°C, 24 h, filtration	53.0	[46]
	Banana juice	150°C, 4 h, centrifugation	9.0	[66]
Microwave treatment	Eggshell Membrane	400°C, 2h, ash added to 1 M NaOH, microwave 5 min, diluted, centrifugation, dialysis.	14	[47]
	Table sugar	150°C, 3 min, NH_3 added, microwave 3 min, 120°C, filtration	2.5	[48]
	Naked oats	pyrolysis, 400°C, 2 h, microwave 12 min, 700 W	3.0	[49]
	Flour	microwave 180°C, 20 min, centrifugation, vacuum-dried	5.4	[50]
	Sesame seeds	800 W, 10−15 min, ground, centrifugation	8.0	[51]
	Rose flowers	dried, microwave, centrifugation, dialysis	13.5	[52]

TABLE 1.1   *(Continued)*

Synthetic method	Precursors	Procedures	QY (%)	Ref.
Pyrolysis	Pigeon feathers		24.9	
	Egg white	300°C, 3 h, centrifugation, dialysis	17.5	[24]
	Egg yolk		16.3	
	Manure		33.5	
	Peanut shells	400°C, 4 h, ground, sieved, 100 mesh, sonication filtration, dialysis	10.6	[36]
	Citrus Limetta Organic waste	190°C, 20 min	63.3	[54]
	Watermelon peel	220°C, 2 h, air, centrifugation, dialysis	7.1	[55]
	Lychee seeds	300°C, 2 h	10.6	[56]
	Black soya beans	200°C, 4 h, centrifugation, dialysis, freeze-dried	38.7	[57]
	Carica papaya Waste pulp	460 K	23.7	[58]
Extraction	Mature vinegar	rotary evaporation, filtration, purified with macroporous resin	5.1	[59]
	Beer	Rotary evaporation, 0.22 µm membrane filtration, Sephadex G-25 gel filtration chromatography	7.39	[60]
Acid oxidation	Lignin	sonication 1 h, 90°C, 8–24 h, dialysis	13.0	[61]
	Manilkara zapota Fruits	freeze-dried, mixed with H_2SO_4, sonication, 100°C, 1 h, dialysis	7.9	[62]
Plasma treatment	Chicken egg	120 W, 3 min, centrifugation, filtration; 250°C, 2 h, centrifugation, filtration	6.0	[63]
Calcination	Green tea	300°C, 2 h, centrifugation	4.3	[64]

TABLE 1.1 *(Continued)*

Synthetic method	Precursors	Procedures	QY (%)	Ref.
Liquid nitrogen-assisted synthesis	Blueberry	Powdered using N_2, sonication, centrifugation	-	[65]
Carbonization	Waste frying oil	100°C in 1 mL of H_2SO_4 for 5 min, pH adjusted by NaOH, dialysis	23.2	[28]

1.2.2 PROPERTIES

CDs exhibited severe complexity in the physicochemical properties caused by their diverse sizes, crystallinity levels, and chemical composition. The structures and components of CDs determine their miscellaneous properties. CDs primarily consist of C, H, N, and O elements, which are present in the form of various functional groups and impart excellent solubility in water and biocompatibility with scope for further functionalization. CDs are suitable for chemical modification and surface inactivation using various biological, polymeric, volatile, or biological materials. By surface passivation, the fluorescence properties and physical properties of CDs are enhanced.

CDs have features such as good conductivity, gentle chemical structure, and photochemical stability. The advantage of green precursors over chemical entities is that most of these methods do not require distinct reactants for doping, post-modifications, or surface passivation because of the presence of various carbohydrates, protein, and other biomolecules which provide self-passivation to green CDs. This section briefly labels the key properties of green CDs.

1.2.2.1 DISPERSIBILITY

Dispersibility is the basic condition for CD solution processing and application. However, this property is strongly dependent on the CD surface chemistry. Most CDs are hydrophilic with good solubility because of the oxygen-containing functional groups derived from precursors or generated during synthesis.[67] Hydrophobic CDs are prepared by the chemical carbonization of hydrophobic organic precursors. Hydrophilicity and dispersibility are strongly influenced by potential CD applications. For example, hydrophibic CDs have been used as a fluorescent probe of image tumors, while hydrophobic CDs have been used as catalysts in many organic reactions.[68]

1.2.2.2   QUANTUM YIELD

QY is the quantitative expression of the fluorescence efficiency. In the early stages, CDs have low QYs due to the reduction of radiative recombination, which is caused by unstable surface defects. The QY of CDs was found to be lower than 1% when they were first calculated.[6] Therefore, several types of research have been devoted for improving the QYs of CDs. To enhance the QYs, there have been several basic studies such as changing reaction approaches, surface modification, and heteroatom doping were promoted.[69] For example, Liu et al. examined the impact of temperature on the formation of carbon-rich photoluminescent polymer nanodots (PPNDs) by HTT of grass. They noted that increasing reaction temperature from 150 to 200°C results in a decrease in size from 22 to 2 nm and a rise in QY from 2.5 to 6.2%.[70]

1.2.2.3   CYTOTOXICITY AND BIOCOMPATIBILITY

Both of them are like facets of a coin. Biocompatible materials do not produce a toxic or immunological response when exposed to the body or bodily fluids. Most of the CDs from the green precursors exhibit low toxicity at moderate concentrations (<10 µg/mL).[71] Low CD toxicity is the observable highlight to fruitful medical application. In some cases, CD toxicity increases significantly in the presence of an external modification. Havrdova et al. synthesized CDs from candle soot, modified with pristine (negative charge), polyethyleneglycol (neutral charge), and polyethylenimine (positive charge). The CDs in vitro toxicity was studied on standard mouse fibroblasts (NIH/3T3). Among this, neutral CDs are the most promising candidate for biological applications because they do not induce any abnormalities in cell morphology, intracellular trafficking, and cell cycle up to concentrations of 300 µg/mL.[72]

1.2.2.4 OPTICAL PROPERTIES

1.2.2.4.1 Absorbance

CDs generally have optical absorption in UV region, while have lower absorption intensity in the visible and near infrared (IR) region. The absorption located in 230–270 nm is generally attributed to π–π* transition of C = C and the shoulder peak extending into the visible absorption region located around 300–330 nm is attributed to n–π* transition of C = O.[73] Recent studies have shown that the absorbance of CDs can be redshifted after specific surface modification. The surface passivation, hybridization, functional groups, size, and structure affect the optical absorption characteristics of CDs.[74]

1.2.2.4.2 Photoluminescence (PL)

The classic signature of quantum confinement is the observation of size-dependent optical absorption. From a fundamental and an application viewpoint, PL is one of the most interesting features of CDs. In general, one uniform feature of the PL for CDs is the distinct dependence of the emission wavelength and intensity. Different possible explanations for this including size dependence, surface defects, surface states, degree of oxidation, etc. The variation of particle size and PL emission can be reflected from the broad and excitation-dependent PL emission spectrum.[75]

The most common color emitted from organic CDs is blue but emissions range from deep ultraviolet (UV) to near-IR. Sahu et al. prepared highly fluorescent blue colored CDs from orange juice with excitation wavelength of 405 nm. Green luminescent water-soluble CDs with an average size of 3 nm and QY of 8.95% on excitation at a wavelength of 360 nm were synthesized by simply heating banana juice at 150°C.[76] Furthermore, yellow and red fluorescence from organic CDs has also been reported.[62,77] Most PL spectra are symmetrical and broad due to the large size distribution of CDs. The fluorescence property can be determined by the π states of the sp^2 sites. The band gap and radiative recombination of electron–hole pairs facilitate the behavior of fluorescence. The σ and σ* states of sp³ hybridization also affect electron–hole recombination and fluorescence behavior.[78]

1.2.2.4.3 Up-Conversion Photoluminescence (UCPL)

Recent studies have also demonstrated that some CDs have UCPL emission feature in addition to the conventional PL emission. The UCPL emission is contrary to normal PL in which the emission wavelength has less energy than the excitation wavelength (anti-Stokes type emission).[79] UCPL of CQDs opens up new opportunities for this cell imaging with two-photon luminescence microscopy for applications in bioscience and energy technology as well as for the design of highly efficient catalyst.

However, some studies report that this upward conversion may be anartifact or PL was an artificial up-conversion emission because it is excited by a second-order variation of light from a harmonic wavelength. After adding a filter to prevent long-wavelength excitation, no upward transition was detected. Consequently, caution must be exercised when interpreting up-conversion emission.[80] It is necessary to eliminate the normal fluorescence and measure the excitation intensity dependence of the fluorescence while observing UCPL. Up-conversion fluorescence behavior was observed in many other green CDs.[25]

1.2.2.5   STRUCTURAL PROPERTIES

CDs typically consist of sp²/sp³ carbon and oxygen. When the CDs are created from heteroatom-rich biomass materials (like protein, hair, onion, etc.) they also contain sulfur and nitrogen groups.[81] The CDs surface was connected to chemical groups, for example, amino groups, oxygen groups, etc. The morphology and size distribution of CDs characterized by transmission electron microscopy (TEM) are spherical in morphology and are well dispersed (Fig. 1.3). Due to disordered carbon atoms, powder X-ray diffraction (XRD) of CDs usually shows an amorphous form. The PXRD spectra of green CDs typically show a broad diffraction peak in the 2θ range 20–25° and the interlayer spacing (d) is in between 0.31 and 0.38 nm. The average particle size in most green CDs is <10 nm with spherical shape.[82–83] The atomic force microscopy (AFM) showed thickness of about 1.5 nm corresponding to 3–5 graphene sheets with a high degree of graphitization as determined by the Raman spectrum. Likewise, D and G bands at ~1372 and ~1638 cm^{-1}, respectively, in the Raman spectrum of CDs confirm their carbonaceous architecture. The presence of relatively high intensity of D band over G band in spectrum also indicates the increased degree of localized sp^3 defects within sp^2 clusters.[84]

The surface functional groups and element states of CDs were characterized by X-ray photoelectron spectroscopy (XPS) and Fourier transform infrared (FT-IR). The XPS analysis of CDs demonstrates that the CDs are mainly composed of carbon and oxygen elements. A classic example of self-passivated CDs is the use of onion peels as a carbon source. The XPS spectrum shows the presence of nitrogen, phosphorus, and sulfur in CDs without any additional reagent. In FT-IR spectrum of CDs, the peaks at ~3426 and ~3151 cm^{-1} are attributed to the O–H and C–H stretching vibrations, the peaks at about ~1634 and ~1400 cm^{-1} indicated the existence of COO$^-$, and the peaks at ~1114 and ~1165 cm^{-1} are attributed to C–O stretching vibrations and the prominent peak at ~952 cm^{-1} is ascribed to the O–H bending vibrations. The functional groups identified by the FT-IR are in good agreement with XPS. ^{13}C nuclear magnetic resonance (NMR) spectra can reveal surface functional group information and also the sp^3/sp^2 hybrid carbon structures of CDs. The two peaks belong to C = C sp^2 carbon (138 ppm) and carboxylic/carbonyl carbon (170–180 ppm), respectively. The ^{13}C NMR spectrum of CDs derived from candle soot by oxidation treatment has three characteristic peaks, which belong to the end C = C bond (114.3 ppm), the internal C = C bond (137.6 ppm), and the C = O bond (173.6 ppm), respectively.[85–88]

Since the invention of CDs began in 2004, many simple, low cost, and efficient routes have been developed for the integration of CDs. Over the past decade, several CDs of different sizes and properties have been successfully produced. The design ability of the green synthesized CDs with surface functional groups affords extensive property and tunability. A vital sustainability-relevant function of CDs has to do with the chemistry of CDs fabrication. CDs have proven benefits over conventional semiconducting quantum dots and organic fluorophores, resulting from their easy synthesis, low cytotoxicity, and superior optical properties. The carbon resources can come from meal wastes, animal wastes, and plant wastes to prevent pollutants, reduce the synthetic cost, and open up new avenues for waste control and at the same time, using waste material provides the opportunity for big-scale preparation of CDs. The properties of CDs making them a perfect platform for several important environmental, organic, and energy-associated applications. The obvious benefits offered by green chemistry have impressed nice efforts within the synthesis and applications of green CDs.

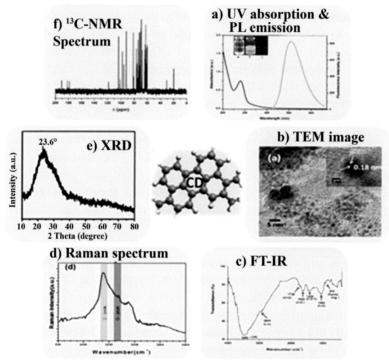

FIGURE 1.3 (a) UV absorption and PL emission spectra of CDs from aloe (b) TEM image of CDs from orange juice (c) FTIR and f) 13C-NMR spectra of CDs from banana juice (d) Raman spectrum of CDs from *citrus limetta* (e) XRD pattern of CDs from waste frying oil.

Sources: a: Reprinted with permission from Ref. [27].© 2015 American Chemical Society. b: Reprinted with permission from Ref. [41]. © 1996 Royal Chemical Society. c: Reprinted with permission from Ref. [81]. © 2011 Royal Chemical Society. d: Reprinted with permission from Ref. [54]. © 2019 American Chemical Society. e: Reprinted with permission from Ref. [28]. © 2014 Elsevier.

1.3   CARBON NANOTUBES

CNTs are nano-architectured allotropes of carbon which are few nanometers in diameter and form one-dimensional structures because the length is orders of magnitude larger than the diameter. Based on the way in which carbon nanotubes are rolled, they are divided into SWCNTs and MWCNTs (Fig. 1.4). SWNTs are formed when single graphene sheets are rolled in the form of a tube whereas MCWNTs are the result of multiple rolled single graphene sheets rolled up to form a tube, while the MWCNTs are the result of concentric tubes of graphene. Due to some unusual properties like high degree of stiffness, a large length-to-diameter ratio and exceptional resilience, CNTs are used for a variety of applications.[89] The report of Sumio Iijima in 1991 introduced CNTs into the scientific community. It was followed by individual publications on the methods to obtain SWNTs by Iijima and Donald Bethune in 1993. All fields of sciences welcomed CNTs with both hands and later it was the era of CNTs.

Several fossil fuel related carbon sources like acetylene, benzene, xylene, toluene, etc. have been used to synthesize CNTs.[90–92] These precursors are not renewable and may vanish with time in the near future. Not only the crisis for fossil fuel-associated carbon precursors but also the destruction to nature highlight the need for natural precursors. Lot of naturally occurring sources are emerging as a new hope with an immense potential of efficiency to synthesize CNTs. Environmentally benign and those synthesis methods of CNTs following principles of green chemistry is of enormous importance. In this section, general methods of synthesis of CNTs use of natural precursors, natural catalysts employed, and the properties are discussed in detail.

(a)   **(b)**

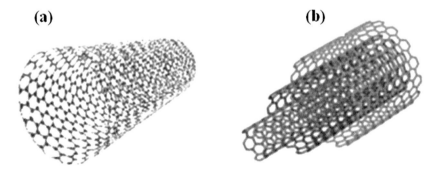

FIGURE 1.4   Representation of (a) single-walled carbon nanotube (SWCNT) and (b) multi-walled carbon nanotube (MWCNT).

1.3.1 SYNTHESIS OF CARBON NANOTUBES

Various ways can be used for the fabrication of carbon nanotubes with required properties for a desired application. The main synthesis methods which produce CNTs with minimum chemical and structural defects are described below.

1.3.1.1 ARC DISCHARGE METHOD

This was the foremost and familiar method for the synthesis of CNTs. In this method, an electric arc discharge is produced between a cathode and graphite anode in a steel chamber with an inert gas. The chamber containing vaporized carbon molecules and catalysts is heated to 4000 K under applied pressure. During this process, a part of the vaporized carbon solidifies as hard cylindrical deposit on the tip of cathode and the remaining part of carbon condenses forming cathode soot and chamber soot. CNTs are formed by cathode soot and anode soot and the selection of inert gas and metallic catalyst added decides whether the CNTs formed are SWCNTs or MWCNTs.[93] Even though this method facilitates the high yield of CNTs, slight control over the orientation of formed CNTs is a major drawback, which eventually affect their activity.

1.3.1.2 LASER ABLATION METHOD

The only difference between laser ablation technique and arc discharge method is in the input energy sources. In this method, a laser acts as an input energy source. A quartz tube containing graphite block is initially heated in a furnace at high temperature using a laser source in the presence of metal particles as catalyst. Argon, which is maintained during the course of reaction carries away the vaporized carbon that finally condenses on the cooler walls of quartz. Studies shows that laser pulse power could influence the diameter and yield of CNTs.[94,95] High yield and relatively low metallic impurity are the advantages of this method. A major limitation of this method is that the synthesized nanotubes may not be straight on a regular basis and have branching to some extent.

1.3.1.3   CHEMICAL VAPOR DEPOSITION (CVD) METHOD

By using CVD approach CNTs can be grown on a variety of materials. This is a widely used method for the synthesis of CNTs by using natural precursors as carbon source. The characteristic growth mechanism of nanotubes in the process of CVD involves the catalyzed dissociation of hydrocarbon molecules by the transition metal and saturation of carbon atoms in the metal nanoparticle. This method can be classified into different types, including catalytic chemical vapor deposition (CCVD)[96] plasma-enhanced chemical vapor deposition (PECVD)[97] microwave plasma-enhanced chemical vapor deposition (MPECVD)[98] and oxygen-assisted CVD. In this technique, catalyst is implanted in a ceramic boat that is planted into a quartz tube. The reaction mixture containing a source of hydrocarbon is passed through a catalyst bed at high temperature and then cooled to room temperature. This technique allows CNTs to grow in an orderly manner with nanostructures designed specifically for some applications. The generation of tubular carbon solids with sp^2 structure occurs when the metal particle precipitates out carbon. Working conditions such as pressure, temperature, concentration, nature of support, and time of reaction have great impact on the characteristics of CNTs.[99,100]

1.3.1.4   SPRAY PYROLYSIS METHOD

This is an advanced form of chemical vapor deposition method for the production of CNTs on different surfaces. In this method, the system is sealed completely to prevent leaking and consists of thermolyne single zone split tube furnace with a quartz tube. Samples are placed about nine inches from the center of tube furnace inside quartz tube. Pure argon is passed through the reactor and the furnace is heated. At a particular temperature and a specific flow rate of argon, the reaction mixture was injected rapidly. CNTs are then obtained in the form of powder after successive chemical treatment. This technique is superior to CVD method since the reaction mixture can be continuously introduced with adequate control into the zone of reaction thereby making the synthesis of MWCNTs continuous and cheaper.[101]

1.3.2 SYNTHESIS OF CNTS USING NATURAL PRECURSORS

Use of natural precursors for the synthesis of CNTs is of immense importance in the current scenario. Green alternatives to fossil fuel-related precursors are safe, environment friendly, renewable, and inexpensive. Several green precursors are reported for the successive production of CNTs.

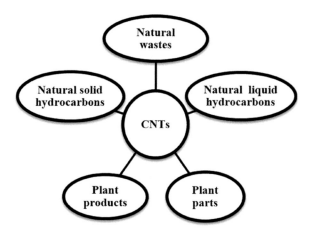

FIGURE 1.5   Type of natural precursors used for the synthesis of CNTs.

Some of the main natural precursors used for the synthesis of CNTs are solid natural hydrocarbon precursors (camphor), liquid natural hydrocarbon precursors (turpentine oil, eucalyptus, palm oil, neem oil, sunflower oil, jatropha-derived bio-diesel, castor oil, sesame oil, camphor oil, tea tree extract) (Fig. 1.5). In addition to these, several plant parts and plant products were also used for the synthesis of carbon nanotubes. The type of precursor, catalyst, and reaction temperature are found to be the most important parameters determining the quantity and quality of nanotube formed. High temperature prefers the formation of SWNTs than MWNTs. Also as the temperature increases the yield of product increases due to the complete cracking of carbon containing compounds. The synthesis method adopted for most of these are either CVD or spray pyrolysis method. A table summarizing the source, type of CNT formed, catalyst used and reaction conditions are listed below (Table 1.2).

TABLE 1.2 Some of the Natural Precursors used for CNT Synthesis, Type of CNTs Formed, Condition, and Catalyst for CNT formation.

Natural source	Type of CNTs formed	Condition and catalyst	Ref.
Turpentine oil	MWCNTs	700°C, ferrocene	[102]
	SWCNTs	850°C, zeolite	[103]
	N doped CNTs	700°C, ferrocene	[104]
Eucalyptus oil	SWCNTs	850°C, silica–zeolite support impregnated with Fe/Co catalyst	[105]
Palm oil	SWCNTs	750°C	[106]
Neem oil	MWCNTs	825°C, ferrocene	[107]
Sunflower oil	N-doped MWCNTs	825°C, NH_3 and ferrocene	[108]
Biodiesel	MWCNTs and N-doped CNTs	850°C (800°C), acetonitrile and ferrocene	[109]
Castor oil			[110]
Sesame oil	MWCNTs	850°C, ferrocene	[111]
Camphor oil	MWCNTs	800°C, SiO_2 substrate	[112]
Camphor	MWCNTs	750–850°C, porous silicon substrate	[113]
Tea tree extract	SWCNTs and MWCNTs	800–1050°C, ferrocene	[114]
	ACNTs	850°C, ferrocene	
Chicken feather	N-doped CNTs	650°C	[115]
Bamboo charcoal	MWCNTs	1200–1400°C	[116]
Maize	Coiled CNTs	1000°C, Ni	[117]
Sugarcane	MWCNTs	500–600°C	[118]
Calotropis latex	CNTs	900°C,Ni	[119]

1.3.3 NATURAL RESOURCES AS CATALYST

The metal catalysts used in the synthesis of CNTs cause serious hazards to living organisms,[120] since it remains as such in the grown CNTs. The catalyst particle can enter into the human body during medical applications. Also in agricultural fields, there are chances for this catalyst to reach plant body resulting in cell division inhibition.[121] The production of CNTs using metal catalyst requires expensive instrumentation and high temperature. All these circumstances points out the need for the development of cheap and nonmetallic catalyst for the safe application in different fields. As a result of intense research in this area, scientists have developed some green approaches for the same. Catalyst-free carbon nanotube synthesis, use of green plant extracts, and naturally available resources as catalysts are some of the best initiative steps taken for the sustainable development.

1.3.3.1   GREEN PLANT EXTRACT AS CATALYST

Plant extracts of neem (*Azadirachta indica*), walnut (*Juglans regia*), garden grass (*Cynodon dactylon*), and rose (Rosa) were used as catalyst for the CNT synthesis. Comparatively high yield than that of other reported metal catalyst, requirement of inexpensive systems for synthesis, and low growth temperature were the main benefits obtained using nontoxic plant extracts.[122]

1.3.3.2   NATURAL MATERIALS AS CATALYST

All natural materials including minerals which contain metals or metal oxides in trace amounts are ideal for the use as catalysts in the synthesis of CNTs. The minerals cannot be classified as a renewable type but they are abundant, inexpensive and are apt to be used as natural catalyst for the preparation of CNTs. Volcanic lava rock (trachybasalt and alkali basalt),[123] garnet (beach sand),[124] bentonite,[125] and red soil (terra rosa)[126] are some of the minerals used for the CNT synthesis. All these examples employed either acetylene or natural gas as carbon precursors. Compared to toxic metal catalyst-mediated CNT synthesis, this approach is environment friendly as it enables the production of CNTs in large scale at low cost. A well-crystallized structure and easy separation are the further plus points of adopting this method.

1.3.3.3   CATALYST-FREE SYNTHESIS OF CNTS

In some cases, the precursor used for the synthesis itself contains certain minerals which can act as catalyst. These minerals are capable of adjusting the size of pores, modifying pyrolytic, and process thereby helping in the growth of CNTs. CNTs were successfully synthesized from bamboo charcoal,[116] sunflower seed hulls and sago,[127] black Jews ear fungus, and black sesame seeds[128] without any external catalyst. The minerals inherently present in these precursors promotes the growth of CNTs in an aligned manner. The use of catalyst-free synthetic precursors does not make any harm on living organisms and also promotes harmless.

1.3.4   PROPERTIES OF CNTS

CNTs are beehive-shaped tubes comprising only of carbon and having a thickness of approximately 1/50,000[th] of human hair. The ratio between length and diameter of graphene sheets, which are rolled into tube is in such a way that they have unique one-dimensional structures. The exceptional properties such as high conductivity, mechanical strength, elasticity, chemical, thermal, and structural stability make CNTs an excellent candidate for a wide range of applications. The properties of CNTs are dependent on the diameter, length, and morphology of tube.[129–131] SWCNTs and MWCNTs differ from each other in various aspects. There are three different forms of SWCNTs such as armchair, chiral, and zigzag, depending on the way in which it is wrapped to form a cylinder (Table 1.3).

TABLE 1.3   Comparison Between SWCNT and MWCNT.

SWCNT	MWCNT
Graphene in single layer	Graphene in multiple layers
Synthesis requires presence of catalyst	Catalyst is not necessary
Difficult to produce in bulk since control is not achieved	Bulk synthesis is easy
Poor purity	High purity
More chance of defect during functionalization	Less chance of defect during functionalization
Easiness in characterization and evaluation	Complex structure
More easiness in twisting	It cannot be easily twisted

The high electrical conductivity of CNTs is due to high electron transfer rate over its sidewall. The curvature of sidewall and structural defects account for the chemical reactivity of CNTs. Increasing the curvature of the sidewall increases the exohedral chemical reactivity at the convex surface of CNTs. This is mainly attributed to the distortions of planar C–C bonds of sp^2 hybridization and misalignment of p orbitals.[132] The available surface area and porous nature of nanotube determines the adsorption properties of CNTs. The higher condensation pressures and minimum heat of adsorption than graphite make CNTs as a better adsorbent.[133] The tube ends and sidewall of CNTs contribute to their EC behavior. Another important property of CNT is their elasticity. When exposed to compressive forces, it can bend and twist without damaging the nanotube and will return to its original state after removing the force. But defects in structure of nanotube can weaken its strength. The modulus of elasticity determines the elasticity in both SWCNTs and MWCNTs.

Carbon nanotubes are an inevitable part of technology with a huge potential of future applications. Green approach is the only answer to solve the dilemma of fossil fuel related source and the pollution it cause to nature. The use of natural resource as precursor and catalyst will not only minimize the cost and utilization of restricted fossil fuels but will also help us to take care of our environment in a more benign way. In this effort, the various synthesis methods employed for the preparation of CNTs involved the common natural precursors used, different green approaches adopted for replacing toxic metal catalysts, and the general properties of CNTs are discussed. The study of properties and applications of CNTs from natural sources is still in progress. Therefore it can be concluded that the CNTs synthesized from green precursors have a very high potential for future applications.

1.4 GRAPHENE/GRAPHENE OXIDE/REDUCED GRAPHENE OXIDE

Graphene is a well-known allotrope of carbon.[134] It was discovered by Andre Geim and Kostya Novoselov by scotch tape method in 2004, and they won Nobel prize for this discovery in 2010.[135] Graphene has attracted an incredible interest in researchers in the modern-day generation of emerging nanochemistry field due to its optical, electronic, thermal, and mechanical properties.[136] Because of the unique physicochemical properties of graphene, it has many applications in science and technology. Graphene is the building block of other important allotropes because zero-dimensional fullerenes are the wrapped form of graphene, nanotubes (1D) as the rolled form of graphene and stacked form of graphene gives the 3D structured graphite.[137] GO is a biocompatible carbon-based nanomaterial. It is an alternative material for graphene due to the similarities in properties. GO is a layered material with a large specific surface area and graphite is the main precursor for the synthesis of GO. Hydroxyl, carbonyl groups, carboxylic, and epoxy group are major groups present in GO[147] and all the groups present are hydrophilic in nature. Due to the presence of these groups, GO is easily dissolved in water. Here we discuss the different green routes for the synthesis of graphene and its oxides.

1.4.1   GENERAL CHARACTERISTICS OF GRAPHENE, GO, AND rGO

The single carbon layer of graphite is called graphene. The main precursor for the synthesis of graphene is GO. Complete reduction of all functional groups (hydroxyl, epoxy, etc.) present in GO results in high-quality graphene. But a partial reduction of GO leads to the formation of reduced graphene oxide (rGO) (Fig. 1.6). The rGO sheets are generally considered as one type of the chemically derived graphene.[138] Graphite is the starting material for the synthesis of graphene, GO, and rGO.

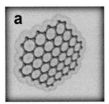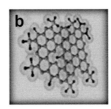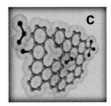

FIGURE 1.6   Schematic representation of (a) graphene, (b) graphene oxide, and (c) reduced graphene oxide.

Large-scale synthesis of high-quality graphene, GO, and rGO in an eco-friendly and inexpensive manner is a big challenge. Lots of chemical methods are used for the production of these carbon-based nanomaterials, like scotch tape method, chemical vapor deposition, vacuum thermal annealing, micro-mechanical exfoliation of graphite, liquid phase exfoliation, epitaxial graphene grown on SiC, chemical intercalation of graphite, CVD/PECVD, epitaxial growth, solvothermal, or chemical reduction of GO and electric arc discharge.[141,142] But intensive energy requirements, high cost, usage of toxic, corrosive chemicals, and formation of hazardous by-products are the major disadvantage of these techniques. Unlike these chemical methods, green or biological methods by using plants and microorganisms do not require any toxic chemicals, high cost, and large amount of energy. Green methods do not produce any hazardous waste and the products usually do not need purification. The green synthesis of graphene, GO, and rGO using plants extract, biomolecules, and microbes provide economic and environmental benefits.[155] Therefore, the green method is an excellent way for the synthesis high-quality graphene, GO, and rGO without any harmful by products.[143,144]

1.4.2 PROPERTIES

1.4.2.1 GRAPHENE

Graphene is an ultrathin material, with a plain sheet of sp^2 hybridized carbon atoms with honey comb like structure. The carbon–carbon distance in graphene is 0.142 nm and the thickness of a single graphene layer is about 0.34 nm.[137] A perfect graphene sheet is very ordered and shows several extraordinary properties including very high electron mobility, strong chemical durability, excellent thermal conductivity, outstanding surface areas, high Young's modulus, and anomalous quantum Hall effect.[139] Due to the two-dimensional structure of graphene, every single atom is exposure to chemical reaction.[155] These special characteristics of graphene opened a new way of extensive range of applications.[144] The band gap of graphene is 0 eV, suggesting that means there is no energy gap between valence band and conduction band. This special property of graphene is mainly applicable in electronic devices.[165] The surface of graphene is highly hydrophobic, since graphene offers an extraordinary support to anchor oxygen-containing functional groups such as, carboxyl, epoxy, and hydroxyl group and forms rGO or GO. Compared to conventional commercial adsorbent, graphene shows better capacity for regeneration, low-temperature modification, and reusability properties. The presence of large and delocalized π- electron system is the reason for the binding of target pollutants.[163]

1.4.2.2   GRAPHENE OXIDE/REDUCED GRAPHENE OXIDE

The structure of GO is similar to graphene, but the GO sheets are thicker than pure graphene sheet because of the presence of carboxylic acid, hydroxyl, carbonyl, and other oxygen-containing functional groups.[141] Due to the presence of these functional groups, GO can be used for the preparation of nanocomposites and GO is highly soluble than graphene. GO and rGO have almost similar properties. The only difference is the decrease in oxygen-containing functional groups in rGO. The UV-vis. absorption of GO shows the peak at 230 nm due to the π-π* transition, but after reduction the absorption peak of rGO at 270 nm undergo red shift due to the removal of oxygen-containing functional groups.[166]

1.4.3   GREEN SYNTHESIS OF GRAPHENE, GRAPHENE OXIDE, AND REDUCED GRAPHENE OXIDE

Two primary approaches for the synthesis of graphene are the top down and bottom up. In top-down rout graphene is synthesized by solution exfoliation of graphite, micromechanical cleavage, and graphitization of SiC, but poor yield is the major disadvantage of this method.[164] Graphene sheets are mainly synthesized by reduction of GO and the GO is reduced to graphene by the help of strong reducing agents. Low cost and high yield are the advantages of this reduction method, but the chemical reduction of GO using toxic and explosive reducing agents like sodium borohydride, hydrazine hydrate, and dimethyl hydrazine cause many environmental issues due to the elimination of hazardous by-products.[142] Therefore, reduction of GO without using hazardous reagents or expensive instruments are very essential for the synthesis of graphene. From the past decade, development of green reducing agents for GO such as plant extract, microbes, biomolecules, etc. are environmentally safe and hazardless reducing agents like sugar, ascorbic acid, heparin, wild carrot roots, aloe vera (AV), vancomycin, *E. fergusoni*, *Escherichia coli, Pseudomonas aeruginosa,* and bovine serum albumin highlight the green approach.[156]

1.4.3.1 REDUCTION USING PLANT EXTRACTS

In general, plants contain many biomolecules, such as polysaccharides amino acids, alkaloids, alcoholic compounds, chelating agents polyphenols, enzymes, vitamins, and proteins. These molecules play an important role in bioreduction.[156] Saikumar Manchala et al. synthesized graphene by the reduction of GO using Eucalyptus polyphenol solution obtained from Eucalyptus bark extract. The polyphenols present in the Eucalyptus bark extract have strong reducing ability, which is responsible for the formation of graphene from GO.[142] Coconut water (*Cocos nucifera L.*) is a natural reducing agent which is nontoxic and environment friendly. Kartick1 et al. synthesized high-quality graphene from GO with the help of coconut water.[155] Porous graphene-like nanosheets (PGNSs) is used as a sensor for hydrogen peroxide. PGNSs were first facitely designed from *ficus-iacertiolia* fruit with active functional groups for the sensing of H_2O_2. This innovative thinking was from Taotao Liang and coworkers.[144] Another green-reducing agent for the synthesis of graphene is from pomegranate juice and anthocyanine is the main content in pomegranate juice. Anthocyanins are water soluble. Due to the hydrophilic nature it can easily convert GO into graphene .This green and economic method for the preparation of graphene nanosheet was introduced by Farnosh Tavakoli et al.[141] Yan Wang and coworkers proposed a method for the reduction of exfoliated GO using green tea solution Due to the presence of aromatic rings of tea polyphenol (TP) is the reason for its reducing capability. The characterization of the obtained graphene confirm the removal of the oxygen-containing functional groups in GO.[148]

One of the interesting route for the preparation of graphene by the deoxygenation of GO using Zante currants (ZC) extract. GO is effectively reduced by ZC. This method shows outstanding reproducibility and this novel synthetic method is proposed by Mohd Zaid Ansari and coworkers.[146] Sangiliyandi Gurunathan et al. reported Ginkgo biloba extract (GbE) as stabilizing and reducing agent for the synthesis of cytocompatible graphene which is an efficient method for the preparation of graphene.[158] EC process is a high efficient and relatively low cost method for the synthesis of GO. The pollution-free environment is the major advantage of this green method. Various graphitic materials such as graphite rode, pencile core, and graphite flakes are used for the synthesis of GO by EC oxidation. Songfeng Pei et al. reported a green method to synthesize high-quality GO by water electrolytic oxidation of graphite.[152] The product of removal of oxygen functional group from GO gives rGO. Lot of green-reducing agents are used for the synthesis of rGO including leaf extracts of *Colocasia esculenta* and *Mesua ferrea Linn*, grape extract *Citrus sinensis* (orange peel).[139]

Gourav Bhattacharya reported a method for the synthesis of rGO from GO using AV as a green-reducing agent. AV extract contains the organic compounds like sugars anthaquinones and polysaccharides, etc. All these have the natural reducing ability.[150] *Azadirachta indica* (AI) which belongs to the family *Meliaceae* and originates in tropical and semitropical regions, alkaloids, flavonoids, and terpenes are the major phytochemicals present in the AI leaves. The presence of these phytochemicals is responsible for the strong reducing tendencies. Using this reducing ability of AI leaves Gnana kumar et al. developed a synthetic route for the preparation of rGO using AI leaves as a reducing agent.[151] Synthesis of rGO from GO by the help of green tea extract is a very simple, efficient, and low cost method introduced by Melvin Jia–Yong Tai and team.[153] Li Gan1 et al. proposed a green approach for the synthesis of rGO sheet. The advantage of this method is the utilization of bagasse and it establishes a waste-to-resource supply chain.[147] For the green synthesis of Pluronic stabilized rGO, ascorbic acid is used as a reducing agent for the reduction of GO to rGO. This is a facile green route was introduced by Cherian R.S et al.[136] Lots of plant extract are used for the synthesis of nanomaterials, among various options, the *Opuntia ficus-indica* (OFI) plant extract is a strong reducer and stabilizer, and using this rGO is synthesized from commercial graphite and also the reaction mixture was placed in a high energy wet ball milling.[145]

1.4.3.2   REDUCTION USING MICROORGANISMS

Various scientific reports are evident for the synthesis of nanoparticles using microorganisms. The mechanism of the reduction reaction depends on the nature of bacterial cells that have the capacity to hydrolyzing the acid group present in GO.[158] An environment friendly green approach for the synthesis of soluble graphene using *Bacillus marisflavi* biomass as a reducing and stabilizing agent under mild conditions in aqueous solution was carried out. The GO formed by this method is applicable for many biomedical applications and this novel idea was suggested by Sangili-yandi Gurunathan et al.[143] Akhavan et al. reported that interactions of the chemically exfoliated GO sheets and *E. coli* bacteria living in mixed-acid fermentation with anaerobic conditions was investigated for the different exposure times of the sheets to the bacteria. The effects of the bacteria and their proliferation on the chemical state, carbon structure, and electrical characteristic of the GO sheets were examined by XPS, Raman spectroscopy, and current–voltage (I–V) measurement, for the different exposure times to the bacteria.[157] Wang et al. reported that *Shewanella* is used for the reduction of GO via external electron transport mediated by c-type cytochromes included heme group and this electron mediators are secreted from *Shewanella*.[162] *Bacillus subtilis* is another type of microorganism used for the reduction of GO and the reduction product is used for the development of a supercapacitor and this excellent idea was developed by Zhang et al.[163]

1.4.3.3   REDUCTION USING BIOMOLECULES

Biomolecules are used as a strong green reducer. Gelatin is a natural reducer, its reducing capability is used for the synthesis of graphene from GO. This interesting idea was proposed by Kunping Liu and coworkers. Gelatin plays an important role for the prevention of aggregation of graphene nanosheet. The synthesized gelatin–GNS have lots of applications. A green and facile method for the preparation of gelatin functionalized graphene nanosheets (gelatin–GNS) was reported by using gelatin as a reducing reagent. Meanwhile, the gelatin also played an important role as a functionalized reagent to prevent the aggregation of the graphene nanosheets.[154] Recently, Zhang et al. suggested that L ascorbic acid is used for the reduction of the GO sheets under mild conditions. This is an excellent method for the synthesis of water-soluble graphene.[149] For the preparation of graphene, vitamin C is used as a secure reductant for the deoxygenation of GO. Vitamin C is a best alternative to the highly toxic reducing agent such as hydrazine and also this reduction reaction is carried out in different organic solvent.[159] The reducing sugars such as fructose and glucose act as mild reducing agents for the synthesis of the graphene nanosheet from GO. In addition to the reducing capability, they also act as capping agent. This was reported by Zhu et al.[160] Another important biomolecule, Melatonin was used for the synthesis of graphene by the reduction of GO with maximum efficiency. This is a biocompatible and very safe bioreducer for the large production of graphene with varity of bio applications.[161]

The green synthesis of graphene, GO, and rGO using plant extract, microorganisms, and biomolecules is efficient, eco-friendly, and cost effective method. Recently, researchers mainly focused on green rout for the synthesis of nanoparticle due its various advantages. Graphite is the main precursor of the synthesis of like graphene, GO, and rGO. The oxidation of graphite gives GO and graphene is a product of strong reduction of GO and partial reduction of GO gives rGO. Reduction using natural-reducing agents happens at mild conditions: atmospheric pressure and room temperature. Considering the tremendous applications of graphene, GO, and rGO, green technology is profoundly promising because of its advantages like nontoxicity, cheap cost, and eco-friendliness. Identification of more efficient green reducers is one of the future challenges of this research field.

1.5 FULLERENE

The third allotrope of carbon fullerene was discovered in 1985 by Harold W. Kroto (University of Sussex, Brighton, England), Robert F. Curl, and Richard E. Smalley (Rice University, Houston, Texas, the United States).[168] In fullerenes each carbon atom is bonded to three other carbon atoms with sp^2 hybridization and is composed of 20 hexagons and 12 pentagons as shown in Figure 1.7.

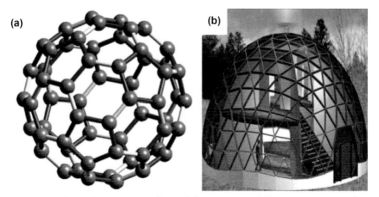

FIGURE 1.7 Schematic representation of the (a) C60 fullerene structure also called buckminsterfullerene and (b) Geodesmic dome built by American architect Richard Buckminster Fuller.

The similarity in the icosahedral symmetry and closed cage structure of fullerene to that of geodesic dome build by American architect Richard Buckminster Fuller, fullerene is called *buckminsterfullerene* . It consists of 60 carbon atoms arranged in the shape of a soccer ball, hence it is also known as buckyball. The discovery of fullerene received appreciation with Nobel Prize in Chemistry in 1996.[169,170]

1.5.1 SYNTHESIS OF FULLERENE

The invention of fullerenes was first reported in 1985 when Smalley and Kroto noticed a carbon cluster containing 60 atoms generating from their laser vaporization apparatus.[171] Since then the research on this new class of allotrope flourished rapidly. The most common synthesis methods of fullerene involve laser vaporization, electric arc heating, laser irradiation of polycyclic hydrocarbons (PAHs), and resistive arc heating. Fullerenes have gained incredible importance in nanoscience due to their outstanding properties and valuable applications. They have been produced from graphite, burning of benzene, combustion of hydrocarbon, and pyrolysis of naphthalene. The high cost and harmful precursors in synthesis is a major concern limiting its application in various fields. Therefore, it is essential to search better alternatives for the production of fullerene using safer and greener sources.

There are only very few reports on the synthesis of fullerene from natural resources. According to the literature review, coal, coke, and camphor were the main natural precursors used for its synthesis.[172–177] Also, there are reports on natural occurrence of fullerene in ancient carbons and suggested to be present in celestial objects such as meteorites. The accidental discovery of natural fullerenes was made by geochemists at Arizona State University at Tempe during the study of a coal-like mineral taken from rock sediments apparently formed during the Precambrian era, more than 600 million years ago.

Coal fines from tailings in coal preparation plants which are regarded as an ongoing pollution hazard were chosen as the precursor for fullerenes. Use of low ash coal as carbon source for fullerene synthesis not only reduces environmental pollution but also transforms into a valuable form with immense potential of applications. The main synthesis method used for the preparation of fullerene from coal was electric arc heating. In electric arc heating, an electric arc was generated between two rods in an inert atmosphere which produces a fluffy soot. The rod should be prepared and it can be changed according to the precursor which is to be used for the synthesis.[178]

Fullerene was also synthesized from camphor (an extract of a tree usually found in Asia) by the vacuum evaporation of toluene extract of camphor soot. In that method, they replaced the tedious chromatographic separation method with hot filament CVD technique.

1.5.2   PROPERTIES OF FULLERENE

The exceptional chemical properties of fullerene are attributed to its electronic structure.[179] Each carbon atom in fullerene is connected to three other carbon atoms resulting in two single bonds and one double bond. Actually there are two types of bonds in C_{60} fullerene namely, a bond shared between a hexagon and pentagon which acts like single bond and that is shared between two hexagons that acts as a double bond. If the fullerene structures have less number of hexagons in it, then it exhibits more sp^3 characteristics such as high reactivity and higher strain. The conjugated system in fullerene is different from that of classical aromatic compounds since there are no replaceable hydrogen atoms that could facilitate substitution reactions. The main two chemical reactions that occur in fullerenes include addition reactions that results in exohedral adduct and redox reactions leading to the formation of salts.[180,181] Fullerenes behave as electron deficient alkenes irrespective of their extreme conjugation. This is evident from their reactions with nucleophiles, homolytic reagents, and free radicals to produce stable adducts.[182]

Fullerenes also have outstanding mechanical properties, high pressure resistance, and capacity to return to original shape even after subjecting to more than 3000 atm. The high bulk modulus indicates that they are harder than steel and diamond.[183] Fullerenes are allotrope of carbon having huge applications especially in medical, engineering, and pharmaceutical fields. But the high cost of synthesis, use of expensive precursors, and catalysts are hindering the practicability of fullerenes. In this perspective, there is high need for the development of inexpensive, renewable, and environment friendly precursors for the synthesis of fullerenes. Unluckily, the green steps taken toward this material are still in its infancy and need to be modified for a better future. The main synthesis methods used for synthesis of fullerene using natural precursors namely coal, coke, and camphor, their advantages, and properties are discussed.

1.6 CARBON NANODIAMONDS

Carbon, one among the nature abundant element can exist in more than one crystalline form called allotropes. Among the different existing carbon nanomaterials (CNs), NDs is considered as the novel member to nano-carbon family. The term "NDs" include materials consisting of nano-sized tetrahedral networks.[184] Diamond nanoparticles usually called as NDs are diamonds with a size below 1 μm.[185] The ND structure is a network comprising of nano-sized tetrahedral frameworks of sp^3 hybridized carbon atoms with small amounts of sp^2 carbons at their surface boundaries.[186] Indifferent to other nanocarbon particles, NDs exhibit unique features like inertness, hardness, conductivity, presence of π-electron network, and easiness in huge surface functionalization that allow for tunability of surface.[186] A wide variety of diamond-based materials at the nanoscale ranging from single diamond clusters to bulk nanocrystalline films have been produced recently.[187,188] Based on molecular arrangements, NDs are classified into nanocrystalline diamond (NCD) particles of size less than 100 nm, ultra-nanocrystalline diamond (UNCD) particles of size less than 10 nm, and diamondoids having particles in the 1–2 nm range. Diamond-like carbon (DLC) films, is a very closely related term is defined by IUPAC[186] as, "are hard, amorphous films with a significant fraction of sp^3-hybridized carbon atoms and which can contain a significant amount of hydrogen."

1.6.1 HISTORY

Nanometer sized diamond are of extra-terrestrial origin, are found in meteorites, protoplanetary nebulae and interstellar dusts, residues of detonation, and in diamond films.[189] It is known that primitive chondritic meteorites contain approximately 1500 ppm of nanometer-sized diamonds, containing isotopically anomalous noble gases like nitrogen and hydrogen.[188] Meteoritic NDs shows potency to provide insights into early solar system formation conditions.[188] ND synthesis was first discovered in 1963 by three scientists K.V. Volkov, V.V. Danilenko, and V.I. Elin in All-Union Research Institute of Technical Physics. This synthesis of NDs was discovered accidentally while studying diamond synthesis by shock compression of nondiamond carbon modifications in blast chambers.[190]

1.6.2 STRUCTURE AND PROPERTIES OF NANODIAMOND

The structure of a ND crystallite consists of a diamond core of sp^3 hybrid carbon atoms that are protected by an amorphous surface layer of carbon atoms in the sp^2 hybridization state (Fig. 1.8). A hybrid layer which exists between the core and the amorphous layer comprises of carbon atoms in both sp^3 and sp^2 hybridization states.[191] The sp^3 organized carbon elements in the core structure of NDs is responsible for its toughness, hardness, chemical stability, high surface areas, biocompatibility, and resistance to harsh conditions.[192,193]

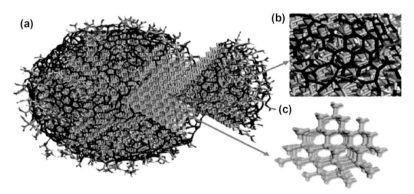

FIGURE 1.8 (a) Model structure of carbon nanodiamond, (b) closer view of nanodiamond surface covered with surface functional groups and sp^2 hybridized carbon, (c) sp^3 hybridized carbon diamond core.

NDs show fascinating optical properties due to the significant band gap (5.49 eV) and high transparency in UV and IR regions.[192,193] The diamond core is responsible for high refractive index of 2.4, which provides high light scattering and hence make them advantageous as nontoxic UV protective nanoadditives in sunscreens and polymer coatings.[194] Optically active imperfection sites present in nandiamond crystal is responsible for many fluorescent applications.[195,196] Stable imperfection sites of nitrogen vacancy can be created in NDs through high energy ion and electron irradiation. In addition to optically active fluorescent defect centers, doping ND core with boron and tritium helps to produce diamonds with applications in the fields of electrical conductivity[196] and bioimaging,[197] respectively.

The significant advantage of NDs on comparison with other nano-carbon materials is the existence of unpaired electrons in the surface layer which allows tunable surface functionalizations, without compromising the diverse properties of the diamond core.[198] ND surface functionalization is of eminent relevance regarding the improvement of particles dispersion, changing their wettability and adhesion, enhancement in catalytic properties. The 1,3-propanediamine functionalized ND shows high solubility in acetone, CH_2Cl_2, NMP, DMF, DMAc, and DMSO. The interaction of ND with sodium bis (2-ethylhexyl) sulfosuccinate resulted in good aqueous dispersions.[197]

1.6.3   SYNTHETIC METHODS FOR NANODIAMOND

Several methods like detonation technique, hydrothermal synthesis, CVD, microwave plasma chemical vapor deposition (MWPCVD), laser bombarding, ion bombarding, autoclave using supercritical fluids, ion irradiation of graphite, chlorination of carbides, ultrasound cavitation, electron irradiation of pure carbon sheets, and EC methods are available for the synthesis of NDs.[199–201] High temperature and pressure conditions are a major requirement for the afore mentioned methods.

1.6.4   GREEN SYNTHESIS OF NANODIAMOND FROM NATURAL PRECURSORS

Xiao et al. reported a novel process of laser ablation in liquid (LAL) as a simple green top-down strategy for the synthesis of NDs from various coals like anthracite, bitumite, and coke.[202] Coal is the most abundant and cheap energy source of carbon and is used for the eco-friendly synthesis of NDs. Anthracite coal consists of micro-sized particles with irregular sizes and shape distributions. While bitumite and coke coals are of similar shapes.

The energy dispersive X-ray spectroscopy (EDS) analysis of anthracite and coke are of purely carbon, while bitumite shows the additional presence of oxygen. The three samples were ablated with a Q-switched Nd:YAG laser device of laser-pulse energy of 200 mJ with a wavelength of 532 nm, having a pulse width of 10 ns, and possessing a repetition frequency of 10 Hz. The different laser irradiations involves a series of colloidal color changes from opaque grayish to dark reddish brown and finally to a transparent yellow, which indicates the formation of various types of products. During all these stages, amorphous carbon particles act as intermediate phase which gets finally transformed into NDs. The green synthesized cubic crystalline NDs show a uniform particle size with a mean size of about 3 nm. These monodisperse colloidal particles are devoid of agglomeration and show a stable bright green fluorescence. The selected area electron diffraction (SAED) pattern shows three rings which corresponds to the (111), (200), and (311) planes of diamonds. The (111) twinning plane of the diamond is often considered as a characteristic of cubic diamonds.

A low-power ultrasound stimulation of low-grade Indian coal mixed with H_2O_2 in an ultrasonic processor (Sonapros; Model: PR-1000 M) at atmospheric pressure and frequency of 20 kHz for 3 h was reported.[203] The filtrate after ultra sonication is neutralized and filtered. During the ultrasound cavitation, very high temperature of 5000 K and pressure of 2000 atm can be attained, which act as initiation point for the formation of NDs.[204,205] Ultrasonicated filtrate containing ND phases show bright blue fluorescence under UV light (at 365 nm) due to size-induced quantum-confinement effect, which finds practical applications in biomedical imaging field. UV-vis. absorption bands appear at 250–350 nm and a shoulder appears around 300 nm due to the $\pi\rightarrow\pi^*$ transition of the aromatic π system and n-π^* transition of C=O bonds, respectively. The broad peak at 26.7° in XRD indicates that the NDs are embedded within an amorphous carbon matrix. TEM analysis gives the planner spacing in the range of 2.0–2.3 Å, which is in good agreement with the lattice planes of various diamond phases including cubic diamond and lonsdaleite. Two Raman peaks at 1600 and 1350 cm^{-1} at positions of G band and D band infer the presence of sp^2 carbons in higher ratio in the product than the characteristic ND sp^3 peaks.

Another novel method for the synthesis of nanaodiamonds from the carbon black precursor was reported.[206] Carbon black is a quasi zero-dimensional structure with very small size. Carbon black is inexpensive and is an ideal source of carbon for the preparation of pure NDs. The procedure involves a long-pulse-width laser irradiation of carbon black in water suspension. The Nd:YAG pulsed laser with a power density of 9 × 106 Wcm^{-2}, having a wavelength length of 1064 nm, frequency 20 Hz, a pulse width of 0.4ms, and an irradiation time of 4 h is used. The laser beam irradiation resulted in vaporization of the surface of carbon black along with a small amount of the surrounding liquid to form bubbles within the water suspension.[207] These species within the bubbles are believed to be subjected to high temperature and pressure conditions, which resulted in the formation of high-pressure-phase structure of nanaodiamonds.[208] The irradiated product is purified by boiling in perchloric acid and passivated by PEG$_{2000N}$. The purified product is heated for 70 h at a temperature of 120°C, is then cooled to room temperature and centrifuged. The SAED pattern of purified product shows rings corresponding to (111), (220), (311), (400), (331), (422), and (511) planes, which indicates diamond type structure. The TEM analysis shows a crystalline diameter of 3.6 nm and an interplanar distance of 0.206 nm concur with (111) plane of cubic diamond. The Raman spectrum shows two intense peaks at 1331 and 1611 cm^{-1}. The former peak corresponds to the first order diamond Raman line which ratifies the presence of diamond particles. The peak at 1611 cm^{-1} was probably due to the paired threefold coordinated defects.

A novel, simple, and efficient extraction of NDs from carbonaceous waste deposits present on the roof of Hawan Kund in Indian Temples was also reported.[209] Metallic impurities of the raw materials were removed by treatment with a mixture of concentrated sulphuric acid (H_2SO_4) and fuming nitric acid (HNO_3). After 3 h of heat treatment of the mixture at 250°C in air, deionised water was added to maintain the pH as 5 and was allowed for sedimentation at room temperature. The decanted phase was dried, crushed into powder, and was then treated with HNO_3/H_2O_2 mixture and heated at 150°C. The final product was washed with deionized water to remove any undesirable carbon from the raw material and the color of the sample changes to grayish black. EDS of extracted NDs show the presence of silicon, oxygen, and carbon. Raman spectrum shows a sharp and intense peak at 1332 cm^{-1} corresponds to high purity of ND core sp^3 character. The TEM image of NDs have a size range of about 4–5 nm. XRD peaks at 43.9° and 75.27° corresponds to (111) and (221) planes of sp^3 diamond core, and a broad diffraction up to 26° corresponds to sp^2 structure. The extracted NDs contain various functional groups on its surface which offers the binding of different biomolecules, which make it suitable for various applications.

Less expensive natural sources can be used as the green precursors for the cost effective synthesis of NDs. NDs are identified as less toxic than all other nanocarbons[210–213] and they emerged as a novel platform for nanoscience and nanotechnology.[214] Due to the nontoxicity, biological compatibility, and luminescent properties, NDs possess a wide range of applications in the fields of biomedical imaging, biology and medicine.[210–213] They also shows considerable potential in the fields of photovoltaics, microelectronics, optoelectronics and biosensing.[211,212] Due to the chemical resistance, hardness, and abrasive nature, they are also used in novel wear resistant polymers, metal coatings[213,214] and lubricant additives.[215] Green synthesized carbon NDs and their rapid functionalization owe wide applications in the current global trend.

1.7   CONCLUSIONS

Broad research endeavors in the course of recent two decades have raised the status of green CNs as one of the most generally utilized classes of nanomaterials. Functionalized green carbon-based nanomaterials such as graphene, fullerenes, carbon nanotube, carbondots, etc. have significant importance because of their unique synthetic and physical properties (thermal and electrical conductivity, high mechanical quality, and optical properties). Wide research undertakes are being made to use these materials for different modern applications including high-quality materials and gadgets. In general, the sum total of what these has been accomplished because of the identification of uniqueness of these materials with their excellent properties encouraged by their nanoscale structures. Besides from the auxiliary representation of allotropes happening in various crystallographic structures, it was gathered that CNs/allotropes incorporate or speak to a lot of materials for the most part with various structure, morphology, and properties, however, containing one significant basic component carbon as the fundamental structure. This corroborates with the notion that carbon is one of the most amazing elements in the periodic table as well as indispensable element in the living world.

Although there are many synthetic routes and chemical precursors for carbon-based nanomaterials, we highlighted the synthesis and properties of green CNs. Despite the numerous researches in these areas, there is still much to investigate the properties in order to bring them into real-world applications. Therefore, here we discussed the main green synthesis and properties of graphene, fullerenes, carbon nanotube, CDs, and carbon nanodiamond toward the development of green nanotechnology.

KEYWORDS

- **green synthesis**
- **properties**
- **carbon dots**
- **carbon nanotubes**
- **graphene**
- **fullerenes**
- **carbon nanodiamonds**

REFERENCES

1. Zhu, S.; Meng, Q.; Wang, L.; Zhang, J.; Song, Y.; Jin, H.; Zhang, K.; Sun, Z.; Wang, H.; Yang, Z. Highly Photoluminescent Carbon Dots for Multicolor Patterning, Sensors, and Bioimaging. *Angew. Chem. Int. Ed.* **2013,** *125,* 4045–4049.
2. Bhunia, S. K.; Saha, A.; Maity, A. R.; Ray, S. C.; Jana, N. R. Carbon Nanoparticle-Based Fluorescent Bioimaging Probes. *Sci. Rep.* **2013,** *3,* 1473.
3. Ma, C. B.; Zhu, Z. T.; Wang, H. X.; Huang.; Zhang, X.; Qi, X.;H. Zhang, H. L.; Zhu, H.; Deng, X.; Peng, Y.; Han, Y.; Zhang, H. A General Solid-State Synthesis of Chemically-Doped Fluorescent Graphene Quantum Dots for Bioimaging and Optoelectronic Applications. *Nanoscale* **2015,** *7,* 10162–10169.
4. Antaris, A. L.; Robinson, J. T,; Yaghi, O. K.; Hong, G.; Diao, S.; Luong, R.; Dai, H. Ultra-Low Doses of Chirality Sorted (6,5) Carbon Nanotubes for Simultaneous Tumor Imaging and Photothermal Therapy. *ACS Nano.* 2013, *7,* 3644–3652.
5. Sun, Y. P.; Zhou, B.; Lin, Y.; Wang, W.; Fernando, K. A. S.; Pathak, P.; Meziani, M. J.; Harruff, B. A.; Wang, X.; Wang, H.; Luo, P. G.; Yang, H.; Kose, M. E.; Chen, B.; Veca, L. M.; Xie, Z. Quantum-Sized Carbon Dots for Bright and Colorful Photoluminescence. *J. Am. Chem.* **2006,** *128,* 7756–7757.
6. Xu, X,; Ray, R.; Gu, Y.; Ploehn, H. J.; Gearheart, L.; Raker, K.; Scrivens, W. A. Electrophoretic Analysis and Purification of Fluorescent Single-Walled Carbon Nanotube Fragments. *J. Am. Chem. Soc.* **2004,** *126,* 12736–12737.

7. Sharma, V.; Tiwari, P.; Mobin, S. M. Sustainable Carbon-Dots: Recent Advances in Green Carbon Dots for Sensing and Bioimaging. *J. Mater. Chem. B.* **2017,** *5,* 8904–8924.

8. Zheng, L.; Chi, Y.; Dong, Y.; Lin, J.; Wang, B. Electrochemiluminescence of Water-Soluble Carbon Nanocrystals Released Electrochemically from Graphite. *J. Am. Chem. Soc.* **2009,** *131,* 4564–4565.

9. Zhuo, S.; Shao, M.; Lee, S. T. Upconversion and Downconversion Fluorescent Graphene Quantum Dots: Ultrasonic Preparation and Photocatalysis. *ACS Nano.* **2012,** *6,* 1059–1064.

10. Lu, J.; Yeo, P. S. E.; Gan, C. K.; Wu, P.; Loh, K. P. Transforming C_{60} Molecules into Graphene Quantum Dots. *Nat. Nanotechnol.* **2011,** *6,* 247–252.

11. Zhou, J.; Booker, C.; Li, R.; Zhou, X.; Sham, T. K.; Sun, X.; Ding, Z. An Electrochemical Avenue O Blue Luminescent Nanocrystals from Multiwalled Carbon Nanotubes (MWCNTs). *J. Am. Chem. Soc.* **2007,** *129,* 744–745.

12. Li, H.; He, X.; Liu, Y.; Huang, H.; Lian, S.; Lee, S.T.; Kang, Z. One-Step Ultrasonic Synthesis of Water-Soluble Carbon Nanoparticles with Excellent Photoluminescent Properties. *Carbon* **2011,** *49,* 605–609.

13. Jiang, J.; He, Y.; Li, S.; Cui, H. Amino Acids as the Source for Producing Carbon Nanodots: Microwave Assisted One-Step Synthesis, Intrinsic Photoluminescence Property and Intense Chemiluminescence Enhancement. *Chem. Commun.* **2012,** *48,* 9634.

14. Dong, Y.; Shao, J.; Chen, C.; Li, H.; Wang, R.; Chi, Y.; Lin, X.; Chen, G. Blue Luminescent Graphene Quantum Dots and Graphene Oxide Prepared by Tuning the Carbonization Degree of Citric Acid. *Carbon* **2012,** *50,* 4738–4743.

15. Bourlinos, A. B.; Karakassides, M. A.; Kouloumpis, A.; Gournis, D.; Bakandritsos, A.; Papagiannouli, L.; Aloukos, P.; Couris, S.; Hola, K.; Zboril, R.; Krysmann, M.; Giannelis, E. P. Synthesis, Characterization and Non-Linear Optical Response of Organophilic Carbon Dots. *Carbon* **2013,** *61,* 640–643.

16. Yang, Z.; Xu, M.; Liu, Y.; He, F.; Gao, F.; Su, Y.; Wei, H.; Zhang, Y. Nitrogen-Doped, Carbon-Rich, Highly Photoluminescent Carbon Dots from Ammonium Citrate. *Nanoscale* **2014,** *6,* 1890–1895.

17. Jaiswal, A.; Ghosh, S. S.; Chattopadhyay, A. One Step Synthesis of C-Dots by Microwave Mediated Caramelization of Poly (Ethylene Glycol). *Chem. Commun.* **2012,** *48,* 407–409.

18. Jiang, H.; Chen, F.; Lagally, M. G.; Denes, F. S. New Strategy for Synthesis and Functionalization of Carbon Nanoparticles. *Langmuir.* **2010,** *26,* 1991–1995.

19. Vedamalai, M.; Periasamy, A. P.; Wang, C. P.; Tseng, Y. T.; Ho, L. C.; Shih, C. C.; Chang, H. T. Carbon Nanodots Prepared from O-Phenylenediamine for Sensing of Cu^{2+} Ions in Cells. *Nanoscale* **2014,** *6,* 13119–13125.

20. Wang, W.; Li, Y.; Cheng, l.; Cao, Z.; Liu, W. Water-Soluble and Phosphorus-Containing Carbon Dots with Strong Green Fluorescence for Cell Labelling. *J. Mater. Chem. B.* **2014,** *2,* 46–48.

21. Zhou, L.; Lin, Y.; Huang, Z.; Ren, J.; Qu, X. Carbon Nanodots as Fluorescence Probes for Rapid, Sensitive, and Label-Free Detection of Hg^{2+} and Biothiols in Complex Matrices. *Chem. Commun.* **2012,** *48,* 1147–1149.

22. Wang, L.; Bi, Y.; Gao, J.; Li, Y.; Ding, H.; Ding, L. Carbon Dots Based Turn-on Fluorescent Probes for the Sensitive Determination of Glyphosate in Environmental Water Samples. *RSC Adv.* **2016,** *6,* 85820–85828.

23. Sun, D.; Ban, R.; Zhang, P. H.; Wu, G. H.; Zhang, J. R.; Zhu, J. J. Hair Fiber as a Precursor for Synthesizing of Sulfur- and Nitrogen-co-Doped Carbon Dots with Tunable Luminescence Properties. *Carbon* **2013,** *64,* 424–434.

24. Ye, Q.; Yan, F.; Luo, Y.; Wang, Y.; Zhou, X.; Chen, L. Formation of N, S-Codoped Fluorescent Arbon Dots from Biomass and their Application for the Selective Detection of Mercury and Iron Ion. *Spectrochim Acta A. Mol. Biomol. Spectrosc.* **2017,** *173,* 854–862.

25. Mehta, V. N.; Jha, S.; Basu, H.; Singhal, R. K.; Kailasa, S. K. One-Step Hydrothermal Approach to Fabricate Carbon Dots from Apple Juice for Imaging of Mycobacterium and Fungal Cells. *Sens. Actuator B-Chem.* **2015,** *213,* 434–443.

26. D'souza, S. L.; Chettiar, S. S.; Koduru, J. R.; Kailasa, S. K. Synthesis of Fluorescent Carbon Dots using Daucus carota subsp. sativus Roots for Mitomycin Drug Delivery. *Optik* **2018,** *158,* 893–900.

27. Xu, H.; Yang, X.; Li, G.; Zhao, C.; Liao, X. Green Synthesis of Fluorescent Carbon Dots for Selective Detection of Tartrazine in Food Samples. *J. Agric. Food Chem.* **2015,** *63,* 6707–6714.

28. Hu, Y.; Yang, J.; Tian, J.; Jia, L.; Yu, J.S. Waste Frying Oil as a Precursor for One-Step Synthesis of Sulfur-Doped Carbon Dots with pH-Sensitive Photoluminescence. *Carbon* **2014,** *77,* 775–782.

29. Zhao, Y.; Li, W.; Zhao, X.; Wang, D. P.; Liu, S. X. Carbon Spheres Obtained via Citric Acid Catalysed Hydrothermal Carbonisation of Cellulose. *Mater. Res. Innov.* **2013,** *17,* 546–551.

30. Zhai, X.; Zhang, P.; Liu, C.; Bai, T.; Li, W.; Dai, L.; Liu, W. Highly Luminescent Carbon Nanodots by Microwave-Assisted Pyrolysis. *Chem. Commun.* **2012,** *48,* 7955.

31. Peng, H.; Travas-Sejdic, J. Simple Aqueous Solution Route to Luminescent Carbogenic Dots from Carbohydrates. *Chem. Mater.* **2009,** *21,* 5563–5565.

32. Liang, Z.; Zeng, L.; Cao, X.; Wang, Q.; Wang, X.; Sun, R. Sustainable Carbon Quantum Dots from Forestry and Agricultural Biomass with Amplified Photoluminescence by Simple NH4OH Passivation. *J. Mater. Chem. C.* **2014,** *2,* 9760–9766.

33. Wang, Y.; Wang, X.; Geng, Z.; Xiong, Y.; Wu, W.; Chen, Y. Electrodeposition of a Carbon Dots/Chitosan Composite Produced by a Simple in situ Method and Electrically Controlled Release of Carbon Dots. *J. Mater. Chem. B.* **2015,** *3,* 7511–7517.

34. Pan, D.; Zhang, J.; Li, Z.; Wu, M. Hydrothermal Route for Cutting Graphene Sheets into Blue-Luminescent Graphene Quantum Dots. *Adv. Mater.* **2010,** *22,* 734–738.

35. Thongpool, V.; Asanithi, P.; Limsuwan, P. Synthesis of Carbon Particles using Laser Ablation in ethanol. *Procedia Eng.* **2012,** *32,* 1054–1060.

36. Ma, X.; Dong, Y.; Sun, H.; Chen, N. Highly Fluorescent Carbon Dots from Peanut Shells as Potential Probes for Copper Ion: The Optimization and Analysis of the Synthetic Process. *Mater. Today Chem.* **2017,** *5,* 1–10.

37. Yao, B.; Huang, H.; Liu, Y.; Kang, Z. Carbon Dots: A Small Conundrum. *Trends Chem.* **2019,** *1,* 235–246.

38. Liu, H.; Ding, J.; Zhang, K.; Ding, L. Construction of Biomass Carbon Dots Based Fluorescence Sensors and their Applications in Chemical and Biological Analysis. *Trends Anal. Chem.* **2019**, *118*, 315–337.

39. Wang, Y.; Hu, A. Carbon Quantum Dots: Synthesis, Properties and Applications. *J. Mater. Chem. C.* **2014**, *2*, 6921.

40. Roy, P.; Periasamy, A. P.; Chuang, C.; Liou, Y. R.; Chen, Y. F.; Joly, J.; Liang, C. T.; Chang, H. T. Plant Leaf-Derived Graphene Quantum Dots and Applications for White LEDs. *New J. Chem.* **2014**, *38*, 4946–4951.

41. Sahu, S.; Behera, B.; Maiti, T. K.; Mohapatra, S. Simple One-Step Synthesis of Highly Luminescent Carbon Dots from Orange Juice: Application as Excellent Bio-Imaging Agents. *Chem. Commun.* **2012**, *48*, 8835.

42. Yu, J.; Song, N.; Zhang, Y.K.; Zhong, S.X.; Wang, A.J.; Chen, J. Green Preparation of Carbon Dots by Jinhua bergamot for Sensitive and Selective Fluorescent Detection of Hg^{2+} and Fe^{3+}. *Sens. Actuator B-Chem.* **2015**, *214*, 29–35.

43. Wang, X.; Feng, Y.; Dong, P.; Huang, J. A Mini Review on Carbon Quantum Dots: Preparation, Properties, and Electrocatalytic Application. *Front Chem.* **2019**, *7*.

44. Mathew, J.; Joy, J.; Kumar, A. S.; Philip, J. White Light Emission by Energy Transfer from Areca Nut Husk Extract Loaded with Carbon Dots Synthesized from the Same Extract. *J. Lumin.* **2019**, *208*, 356–362.

45. Suvarnaphaet, P.; Tiwary, C.S.; Wetcharungsri, J.; Porntheeraphat, S.; Hoonsawat, R.; Ajayan, P. M.; Tang, I. M.; Asanithi, P. Blue Photoluminescent Carbon Nanodots from Limeade. *Mater. Sci. Eng. C.* **2016,** *69*, 914–921.

46. Ren, G.; Tang, M.; Chai, F.; Wu, H. One-Pot Synthesis of Highly Fluorescent Carbon Dots from Spinach and Multipurpose Applications. *Eur. J. Inorg. Chem.* **2018**, *2018*, 153–158.

47. Wang, Q.; Liu, X.; Zhang, L.; Lv, Y. Microwave-Assisted Synthesis of Carbon Nanodots through an Eggshell Membrane and their Fluorescent Application. *Analyst* **2012**, *137*, 5392.

48. Ansi, V.; Renuka, N. Table Sugar Derived Carbon dot – a Naked Eye Sensor for Toxic Pb^{2+} Ions. *Sens. Actuator B-Chem.* **2018**, *264*, 67–75.

49. Shi, L.; Li, X.; Li, Y.; Wen, X.; Li, J.; Choi, M.M.F.; Dong, C.; Shuang, S. Naked Oats-Derived Dual-Emission Carbon Nanodots for Ratiometric Sensing and Cellular Imaging. *Sens. Actuator B-Chem.* **2015,** *210*, 533–541.

50. Qin, X.; Lu, W.; Asiri, A.M.; Al-Youbi, A.O.; Sun, X. Microwave-Assisted Rapid Green Synthesis of Photoluminescent Carbon Nanodots from Flour and their Applications for Sensitive and Selective Detection of Mercury (II) Ions. *Sens. Actuator B-Chem.* **2013**, *184*, 156–162.

51. Roshni, V.; Divya, O. One-Step Microwave-Assisted Green Synthesis of Luminescent n-Doped Carbon Dots from Sesame Seeds for Selective Sensing of fe (III). *Current Sci.* **2017**, *112*, 385.

52. Feng, Y.; Zhong, D.; Miao, H.; Yang, X. Carbon Dots Derived from Rose Flowers for Tetracycline Sensing. *Talanta* **2015**, *140*, 128–133.

53. Chen, B.; Li, F.; Li, S.; Weng, W.; Guo, H.; Guo, T.; Zhang, X.; Chen, Y.; Huang, T.; Hong, X.; You, S.; Lin, Y.; Zeng, K.; Chen, S. Large Scale Synthesis of Photoluminescent Carbon Nanodots and their Application for Bioimaging. *Nanoscale* **2013,** *5*, 1967.

54. Thakur, A.; Devi, P.; Saini, S.; Jain, R.; Sinha, R. K.; Kumar, P. Citrus limetta Organic Waste Recycled Carbon Nanolights: Photoelectro Catalytic, Sensing, and Biomedical Applications. *ACS Sustain. Chem. Eng.* **2018**, *7*, 502–512.

55. Zhou, J.; Sheng, Z.; Han, H.; Zou, M.; Li, C. Facile Synthesis of Fluorescent Carbon Dots using Watermelon Peel as a Carbon Source. *Mater. Lett.* **2012**, *66*, 222–224.

56. Xue, M.; Zou, M.; Zhao, J.; Zhan, Z.; Zhao, S. Green Preparation of Fluorescent Carbon Dots from Lychee Seeds and their Application for the Selective Detection of Methylene Blue and Imaging in Living Cells. *J. Mater. Chem. B.* **2015**, *3*, 6783–6789.

57. Jia, J.; Lin, B.; Gao, Y.; Jiao, Y.; Li, L.; Dong, C.; Shuang, S. Highly Luminescent N-Doped Carbon Dots from Black Soya Beans for Free Radical Scavenging, Fe^{3+} Sensing and Cellular Imaging. *Spectrochim Acta A. Mol. Biomol. Spectrosc.* **2019**, *211*, 363–372.

58. Singh, P. D. L.; Thakur, A.; Kumar, P. Green Synthesis of Glowing Carbon Dots from Carica Papaya Waste Pulp and their Application as a Label-Free Chemo Probe for Chromium Detection in Water. *Sens. Actuator B-Chem.* **2019**, *283*, 363–372.

59. Cao, L.; Song, X.; Song, Y.; Bi, J.; Cong, S.; Yu, C.; Tan, M. Fluorescent Nanoparticles from Mature Vinegar: Their Properties and Interaction with Dopamine. *Food Funct.* **2017**, *8*, 4744–4751.

60. Wang, Z.; Liao, H.; Wu, H.; Wang, B.; Zhao, H.; Tan, M. Fluorescent Carbon Dots from Beer for Breast Cancer Cell Imaging and Drug Delivery. *Anal. Methods* **2015**, *7*, 8911–8917.

61. Myint, A. A.; Rhim, W. K.; Nam, J. M.; Kim, J.; Lee, Y. W. Water-Soluble, Lignin-Derived Carbon Dots with High Fluorescent Emissions and their Applications in Bioimaging. *J. Ind. Eng. Chem.* **2018**, *66*, 387–395.

62. Bhamore, J. R.; Jha, S.; Park, T. J.; Kailasa, S. K. Green Synthesis of Multi-Color Emissive Carbon Dots from Manilkara zapota Fruits for Bioimaging of Bacterial and Fungal Cells. *J. Photochem. Photobiol. B. Biol.* **2019**, *191*, 150–155.

63. Wang, J.; Wang, C. F.; Chen, S.; Amphiphilic Egg-Derived Carbon Dots: Rapid Plasma Fabrication, Pyrolysis Process, and Multicolor Printing Patterns. *Angew. Chem. Int. Ed.* **2012**, *51*, 9297–9301.

64. Hsu, P. C.; Chen, P. C.; Ou, C. M.; Chang, H. Y.; Chang, H. T. Extremely High Inhibition Activity of Photoluminescent Carbon Nanodots Toward Cancer Cells. *J. Mater. Chem. B.* **2013**, *1*, 1774.

65. Aslandaş, A. M.; Balcı, N.; Arık, M.; Şakiroğlu, H.; Onganer, Y.; Meral, K. Liquid Nitrogen-Assisted Synthesis of Fluorescent Carbon Dots from Blueberry and their Performance in Fe^{3+} Detection. [Appl. Surf. Sci. **2015**, *356*, 747–752.

66. De, B., Karak, N. A Green and Facile Approach for the Synthesis of Water Soluble Fluorescent Carbon Dots from Banana Juice. *RSC Adv.* **2013**, *3*, 8286–8290.

67. Bourlinos, A. B.; Stassinopoulos, A.; Anglos, D.; Zboril, R.; Georgakilas, V.; Giannelis, E. P. Photoluminescent Carbogenic Dots. *Chem. Mater.* **2008**, *20*, 4539–4541.

68. Zheng, M.; Ruan, S.; Liu, S.; Sun, T.; Qu, D.; Zhao, H.; Xie, Z.; Gao, H.; Jing, X.; Sun, Z. Self-Targeting Fluorescent Carbon Dots for Diagnosis of Brain Cancer Cells. *ACS Nano.* **2015**, *9*, 11455–11461.

69. Liao, J.; Cheng, Z.; Zhou, L. Nitrogen-Doping Enhanced Fluorescent Carbon Dots: Green Synthesis and their Applications for Bioimaging and Label-Free Detection of Au^{3+} Ions. *ACS Sustain. Chem. Eng.* **2016**, *4*, 3053–3061.

70. Liu, S.; Tian, J.; Wang, L.; Zhang, Y.; Qin, X.; Luo, Y.; Asiri, A. M.; Al-Youbi, A. O.; Sun, X. Hydrothermal Treatment of Grass: A Low-Cost, Green Route to Nitrogen-Doped, Carbon-Rich, Photoluminescent Polymer Nanodots as an Effective Fluorescent Sensing Platform for Label-Free Detection of Cu(II) ions. *Adv. Mater.* **2012**, *24*, 2037–2041.

71. Nurunnabi, M.; Khatun, Z.; Huh, K. M.; Park, S. Y.; Lee, D. Y.; Cho, K. J.; Lee, Y. In vivo Biodistribution and Toxicology of Carboxylated Graphene Quantum Dots. *ACS Nano.* **2013**, *7*, 6858–6867.

72. Havrdova, M.; Hola, K.; Skopalik, J.; Tomankova, K.; Petr, M.; Cepe, K.; Polakova, K.; Tucek, J.; Bourlinos, A. B.; Zboril, R. Toxicity of Carbon Dots – Effect of Surface Functionalization on the Cell Viability, Reactive Oxygen Species Generation and Cell Cycle. *Carbon* **2016**, *99*, 238–248.

73. Arcudi, F.; Đorđević, L.; Prato, M. Rationally Designed Carbon Nanodots Towards Pure White-Light Emission. *Angew. Chem. Int. Ed.* **2017**, *56*, 4170–4173.

74. Hu, S. L.; Niu, K. Y. L; Sun, J.; Yang, J.; Zhao, N. Q.; Du, X. W. One-Step Synthesis of Fluorescent Carbon Nanoparticles by Laser Irradiation. *J. Mater. Chem.* **2009**, *19*, 484–488.

75. Jiang, K.; Wang, Y.; Gao, X.; Cai, C.; Lin, H. Facile, Quick, and Gram-Scale Synthesis of Ultralong-Lifetime Room-Temperature-Phosphorescent Carbon Dots by Microwave Irradiation. *Angew. Chem. Int. Ed.* **2018**, *57*, 6216–6220.

76. De, B.; Karak, N.; A Green and Facile Approach for the Synthesis of Water Soluble Fluorescent Carbon Dots from Banana Juice. *RSC Adv.* **2013**, *3*, 8286.

77. Lin, P.Y.; Hsieh, C. W.; Kung, M. L.; Chu, L. Y.; Huang, H. J.; Chen, H. T.; Wu, D. C.; Kuo, C. H.; Hsieh, S. L.; Hsieh, S. Eco-Friendly Synthesis of Shrimp Egg-Derived Carbon Dots for Fluorescent Bioimaging. *J. Biotechnol.* **2014**, *189*, 114–119.

78. Semeniuk, M.; Yi, Z.; Poursorkhabi, V.; Tjong, J.; Jaffer, S.; Lu, Z. H.; Sain, M. Future Perspectives and Review on Organic Carbon Dots in Electronic Applications. *ACS Nano.* **2019**, *13*, 6224–6255.

79. Jiang, C.; Wu, H.; Song, X.; Ma, X.; Wang, J.; Tan, M. Presence of Photoluminescent Carbon Dots in Nescafe® Original Instant Coffee: Applications to Bioimaging. *Talanta* **2014**, *127*, 68–74.

80. Gan, Z.; Wu, X.; Zhou, G.; Shen, J.; Chu, P. K. Is there Real Upconversion Photoluminescence from Graphene Quantum Dots?. *Adv. Opt. Mater.* **2013**, *1*, 554–558.

81. Liu, R.; Zhang, J.; Gao, M.; Li, Z.; Chen, J.; Wu, D.; Liu, P. A Facile Microwave-Hydrothermal Approach Towards Highly Photoluminescent Carbon Dots from Goose Feathers. *RSC Adv.* **2015**, *5*, 4428–4433.

82. Ahmadian-Fard-Fini, S.; Salavati-Niasari, M.; Ghanbari, D. Hydrothermal Green Synthesis of Magnetic Fe_3O_4-Carbon Dots by Lemon and Grape Fruit Extracts and as a Photoluminescence Sensor for Detecting of E. coli bacteria. *Spectrochim Acta A. Mol. Biomol. Spectrosc.* **2018**, *203*, 481–493.

83. Feng, J.; Wang, W. J.; Hai, X.; Yu, Y. L.; Wang, J. H. Green Preparation of Nitrogen-Doped Carbon Dots Derived from Silkworm Chrysalis for Cell Imaging. *J. Mater. Chem. B.* **2016**, *4*, 387–393.

84. Ray, S. C.; Saha, A.; Jana, N. R.; Sarkar, R. Fluorescent Carbon Nanoparticles: Synthesis, Characterization, and Bioimaging Application. *J. Phys. Chem. C.* **2009**, *113*, 18546–18551.

85. Bankoti, K.; Rameshbabu, A. P.; Datta, S.; Das, B.; Mitra, A.; Dhara, S. Onion derived Carbon Nanodots for Live Cell Imaging and Accelerated Skin Wound Healing. *J. Mater. Chem. B.* **2017**, *5*, 6579–6592.

86. Tian, L.; Ghosh, D.; Chen, W.; Pradhan, S.; Chang, X.; Chen, S. Nanosized Carbon Particles from Natural Gas Soot. *Chem. Mater.* **2009**, *21*, 2803–2809.

87. Zhao, S.; Lan, M.; Zhu, X.; Xue, H.; Ng, T. W.; Meng, X.; Lee, C. S.; Wang, P.; Zhang, W. Green Synthesis of Bifunctional Fluorescent Carbon Dots from Garlic for Cellular Imaging and Free Radical Scavenging. *ACS Appl. Mater. Interfaces.* **2015**, *7*, 17054–17060.

88. Liu, H.; Ye, T.; Mao, C. Fluorescent Carbon Nanoparticles Derived from Candle Soot. *Angew. Chem. Int. Ed.* **2007**, *46*, 6473–6475.

89. Rahman, G.; Najaf, Z.; Mehmood, A.; Bilal, S.; Shah, A.; Mian, S. A.; Ali, G. An Overview of the Recent Progress in the Synthesis and Applications of Carbon Nanotubes. *J. Carbon Res.* **2019**, *5*, 3.

90. Mayne, M.; Grobert, N.; Terrones, M.; Kamalakaran, R.; Rühle, M.; Kroto, H. W. Pyrolytic Production of Aligned Carbon Nanotubes from Homogeneously Dispersed Benzene-Based Aerosols. *Chem. Phys. Lett.* **2001**, *338*, 101–107.

91. Bronikowski, M. J. CVD Growth of Carbon Nanotube Bundle Arrays. *Carbon* **2006**, *44*, 2822–32.

92. Zhang, Z. J.; Wei, B. Q.; Ramanath, G.; Ajayan, P. M. Substrate-Site Selective Growth of Aligned Carbon Nanotubes. *Appl. Phys, Lett.* **2000**, *77*, 3764–3766.

93. Eatemadi, A.; Daraee, H.; Karimkhanloo, H.; Kouhi, M.; Zarghami, N.; Akbarzadeh, A. Abasi, M.; Hanifehpour, Y.; Joo, S. W. Carbon Nanotubes: Properties, Synthesis, Purification, and Medical Applications. *Nanoscale Res. Lett.* **2014**, *9*, 393.

94. José-Yacamán, M.; Miki-Yoshida, M.; Rendon, L.; Santiesteban, J. Catalytic Growth of Carbon Microtubules with Fullerene Structure. *Appl. Phys. Lett.* **1993**, *62*, 202–204.

95. Thess, A.; Lee, R.; Nikolaev, P.; Dai, H. Crystalline Ropes of Metallic Carbon Nanotubes. *Science* **1996**, *273*, 483.

96. Seo, J. W.; Magrez, A.; Milas, M.; Lee, K.; Lukovac, V.; Forro, L. Catalytically Grown Carbon Nanotubes: From synthesis to toxicity. *J. Phys. D. Appl. Phys.* **2007**, *40*, R109.

97. Li, Y.; Mann, D.; Rolandi, M.; Kim, W.; Ural, A.; Hung, S.; Javey, A.; Cao, J.; Wang, D.; Yenilmez, E. Preferential Growth of Semiconducting Single-Walled Carbon Nanotubes by a Plasma Enhanced CVD Method. *Nano Lett.* **2004**, *4*, 317–321.

98. Ebbesen, T.; Ajayan, P. Large-Scale Synthesis of Carbon Nanotubes. *Nature* **1992**, *358*, 220.

99. Paradise, M.; Goswami, T. Carbon Nanotubes—Production and Industrial Applications. *Mater. Des.* **2007**, *28*, 1477–1489.

100. Journet, C.; Bernier, P. Production of Carbon Nanotubes. *Appl. Phys. A. Mater. Sci. Process.* **1998**, *67*, 1–9.

101. Wasel, W.; Kuwana, K.; Reilly, P. T.; Saito, K. Experimental Characterization of the Role of Hydrogen in CVD Synthesis of MWCNTs. *Carbon* **2007,** *45,* 833–838.

102. Afre, R. A.; Soga, T.; Jimbo, T.; Kumar, M.; Ando, Y.; Sharon, M. Growth of Vertically Aligned Carbon Nanotubes on Silicon and Quartz Substrate by Spray Pyrolysis of a Natural Precursor: Turpentineoil. *Chem. Phys. Lett.* **2005,** *414*(1–3), 6–10.

103. Ghosh, P.; Soga, T.; Afre, R. A.; Jimbo, T. Simplified Synthesis of Single-Walled Carbon Nanotubes from a Botanical Hydrocarbon: Turpentine Oil. *J. Alloy Compd.* **2008,** *462,* 289–293.

104. Ghosh, P.; Soga, T.; Ghosh, K.; Afre, R. A.; Jimbo, T.; Ando, Y. Vertically Aligned N- Doped Carbon Nanotubes by Spray Pyrolysis of Turpentine Oil and Pyridine Derivative with Dissolved Ferrocene. *J. Noncryst. Solids* **2008,** *354,* 4101–4106.

105. Ghosh, P.; Afre, R. A.; Soga, T.; Jimbo, T. A Simple Method of Producing Single-Walled Carbon Nanotubes from a Natural Precursor: Eucalyptus Oil. *Mater. Lett.* **2007,** *61,* 3768–3770.

106. Suriani, A. B.; Azira, A. A.; Nik, S. F.; MdNorR, Rusop, M.; Synthesis of Vertically Aligned Carbon Nanotubes using Natural Palm Oil as Carbon Precursor. *Mater. Lett.* **2009,** *63,* 2704–2706.

107. Kumar, R.; Tiwari, R.; Srivastava, O. Scalable Synthesis of Aligned Carbon Nano-Tubes Bundles using Green Natural Precursor: Neem Oil. *Nanoscale Res. Lett.* **2011,** *6,* 92.

108. Kumar, R.; Yadav, R. M.; Awasthi, K.; Tiwari, R. S.; Srivastava, O. N. Effect of Nitrogen Variation on the Synthesis of Vertically Aligned Bamboo-Shaped C–N Nano-Tubes using Sunflower Oil. *Int. J. Nanosci.* **2011,** *10,* 809–813.

109. Kumar, R.; Yadav, R. M.; Awasthi, K.; Shripathi, T.; Sinha, A. S. K.; Tiwari, R. S. Synthesis of Carbon and Carbon–Nitrogen Nanotubes using Green Precursor: Jatropha-Derived Biodiesel. *J. Exp. Nanosci.* **2012,** *8,* 606–620.

110. Awasthi, K.; Kumar, R.; Raghubanshi, H.; Awasthi, S.; Pandey, R.; Singh, D. Synthesis of Nano-Carbon (Nanotubes, Nanofibres, Graphene) Materials. *Bull Mater. Sci.* **2011,** *34,* 607–614.

111. Kumar, R.; Singh, R. K.; Kumar, P.; Dubey, P. K.; Tiwari, R. S.; Srivastava, O. N. Clean and Efficient Synthesis of Graphene Nanosheets and Rectangular Aligned-Carbon Nanotubes Bundles using Green Botanical Hydrocarbon Precursor: Sesame Oil. *Sci. Adv. Mater.* **2014,** *6,* 76–83.

112. Termeh Yousefi, A.; Bagheri, S.; Shinji, K.; Rouhi, J.; Rusop Mahmood, M.; Ikeda, S. Fast Synthesis of Multilayer Carbon Nanotubes from Camphor Oil as an Energy Storage Material. *Biomed. Res. Int.* **2014,** *2014,* 1–6.

113. Kumar, M.; Ando, Y. Single-Wall and Multi-Wall Carbon Nanotubes from Camphor—a Botanical Hydrocarbon. *Diam. Relat. Mater.* **2003,** *12,* 1845–1850.

114. Musso, S.; Porro, S.; Giorcelli, M.; Chiodoni, A.; Ricciardi, C.; Tagliaferro, A. Macroscopic Growth of Carbon Nanotube Mats and their Mechanical Properties. *Carbon* **2007,** *45,* 1133–1136.

115. Gao, L.; Li, R.; Sui, X.; Li, R.; Chen, C.; Chen, Q. Conversion of Chicken Feather Waste to N-Doped Carbon Nanotubes for the Catalytic Reduction of 4-Nitrophenol. *Environ. Sci. Technol.* **2014,** *48,* 10191–10197.

116. Zhu, J.; Jia, F.; Kwong, L.; Leung Ng, D. H.; Tjong, S. C. Synthesis of Multiwalled Carbon Nanotubes from Bamboo Charcoal and the Roles of Minerals on their Growth. *Biomass Bioenergy* **2012**, *36*, 12–19.
117. Shukla, J.; Maldar, N. N.; Sharon, M.; Tripathi, S.; Sharon, M. Synthesis of Carbon Nano Material From Different Parts of Maize using Transition Metal Catalysts. *Der. Chemica. Sinic.* **2012**, *3*(5), 1058–1070.
118. Romanovicz, V.; Berns, B. A.; Carpenter, S. D.; Carpenter, D. Carbon Nanotubes Synthesized using Sugar Cane as a Percursor. *Int. J. Chem. Mol. Nucl. Mater. Metall. Eng.* **2013**, *7*, 665–668.
119. Viswanathan, G.; Bhowmik, S.; Sharon, M. Impact of Temperature and Carrier Gas on the Morphology of Carbon Nanomaterials Obtained from Plant Latex. *Int. J. Eng. Res. Gen. Sci.* **2015**, *2*, 93–100.
120. Panyala, N. R.; Mendez, M. P.; Havel, J.; Silver or Silver Nanoparticles: A Hazardous Threat to the Environment and Human Health. *J. Appl. Biomed.* **2008**, *6*,117–129.
121. Masarovicova, E.; Kralova, K. Metal Nanoparticles and Plants. *Ecol. Chem. Eng. S.* **2013**, *20*, 9–12.
122. Tripathi, N.; Pavelyev, V.; Islam, S. S.; Synthesis of Carbon Nanotubes using Green Plant Extract as Catalyst: Unconventional Concept and its Realization. *Appl. Nanosci.* **2017**, *7*, 557–566.
123. Su, D. S.; Chen, X. W. Natural Lavas as Catalysts for Efficient Production of Carbon Nanotubes and Nanofibers. *Angewandte. Chem.* **2007**, *119*, 1855–1856.
124. Endo, M.; Takeuchi, K.; Kim, Y. A.; Park, K. C.; Ichiki, T.; Hayashi, T.; Fukuyo, T.; Iinou, S.; Su, D. S.; Terrones, M.; Dresselhaus, M. S. Simple Synthesis of Multiwalled Carbon Nanotubes from Natural Resources. *Chem. Sus. Chem.* **2008**, *1*, 820–822.
125. Rinaldi, A.; Zhang, J.; Mizera, J.; Girgsdies, F.; Wang, N.; Hamid, S. B. A.; Schlögl, R.; Su, D. S. Facile Synthesis of Carbon Nanotube/Natural Bentonite Composites as a Stable Catalyst for Styrene Synthesis. *Chem. Commun.* **2008**, *48*, 6528–6530.
126. Su, D. S. The use of Natural Materials in Nanocarbon Synthesis. *Chem. Sus. Chem.* **2009**, *2*, 1009–1020.
127. Zhang, H. Y.; Niu, H. J.; Wang, Y. M.; Wang, C.; Bai X. D.; Wang, S.; Wang, W. A. Simple Method to Prepare Carbon Nanotubes from Sunflower Seed Hulls and Sago and their Application in Supercapacitor. *Pigm. Resin Technol.* **2015**, *44*, 7–12.
128. Zhao, J.; Guo, X.; Gu, L.; Guo, Y.; Feng, F. Growth of Carbon Nanotubes on Natural Organic Precursors by Chemical Vapour Deposition. *Carbon* **2011**, *49*, 2155–2158.
129. Thostenson, E. T.; Ren, Z. F.; Chou, T. W. Advances in the Science and Technology of Carbon Nanotubes and their Composites: A Review. *Compos. Sci. Technol.* **2001**, *61*, 1899–1912.
130. Charlier, J. C. Defects in Carbon Nanotubes. *Acc. Chem. Res.* **2002**, *35*, 1063–1069.
131. Gooding, J. J. Nanostructuring Electrodes with Carbon Nanotubes: A Review on Electrochemistry and Applications for Sensing. *Electrochim. Acta.* **2005**, *5*, 3049–3060.
132. Tan, C. W.; Tan, K. H.; Ong, Y. T.; Mohamed, A. R.; Zein, S. H. S.; Tan, S. H. Energy and Environmental Applications of Carbon Nanotubes. *Environ. Chem. Lett.* **2012**, *10*, 265–273.

133. Masenelli Varlot, K.; McRae, E.; Dupont-Pavlovsky, N. Comparative Adsorption of Simple Molecules on Carbon Nanotubes Dependence of the Adsorption Properties on the Nanotube Morphology. *Appl. Surf Sci.* **2002,** *196*, 209–215.

134. Jahangirian, H.; Ghasemianlemraski, E.; Webster, T. J.; Rafiee-Moghaddam, R.; Abdollahi, Y. A Review of Drug Delivery Systems Based on Nanotechnology and Green Chemistry: Green Nanomedicine. *Int. J. Nanomed.* **2017,** *12*, 2957–2978.

135. Cheong, M. F.; Liu, W. W.; Khe, C. S.; Hidayah, N. M. S.; Lee, H. C.; Teoh ,Y. P.; Foo, K. L.; Voon C. H.; Zaaba N. I.; Adelyn, P. Y. P. Green Synthesis of Reduced Graphene Oxide Decorated with Iron Oxide Nanoparticles using Oolong Tea extract. *AIP Conf. Proc.* **2018,** *2045*, 020031.

136. Cherian, R. S.; Sandeman, S.; Ray, S.; Savina, I. N.; A. J.; M. P.V. Green Synthesis of Pluronic Stabilized Reduced Graphene Oxide: Chemical and Biological Characterization. *Colloids Surf. B. Biointerfaces* **2019,** *179*, 94–106.

137. Notarianni, M.; Liu, J.; Vernon, K.; Motta, N. Synthesis and Applications of Carbon Nanomaterials for Energy Generation and Storage. *Nanoonline* **2017,** *2190*, 4286.

138. Pei, S.; Cheng, H. M. The Reduction of Graphene Oxide. *Carbon* **2012,** *50*, 3210–3228.

139. Aunkor, M. T. H.; Mahbubul, I. M.; Saidur, R.; Metselaar, H. S. C. The Green Reduction of Graphene Oxide. *RSC Adv.* **2016,** *6*, 27807–27828.

140. De, S.; Huang, H. H.;. Kanishka, K. H.; Suzuki, S.; Badam, R.; Yoshimura, M. Ethanol-Assisted Restoration of Graphitic Structure with Simultaneous Thermal Reduction of Graphene Oxide. *JJAP* **2018,** *57*, 08NB03-1-6.

141. Tavakoli, F.; Salavati-Niasari, M.; Badiei, A.; Mohandes, F. Green Synthesis and Characterization of Graphene Nanosheets. *Mater. Res. Bull.* **2015,** *63*, 51–57.

142. Manchala, S.; Tandava, V. S. R. K.; Jampaiah, D.; Bhargava, S. K.; Shanker, V. Novel and Highly Efficient Strategy for the Green Synthesis of Soluble Graphene by Aqueous Polyphenol Extracts of Eucalyptus Bark and its Applications in High-Performance Supercapacitors. *ACS Sustainable Chem. Eng.* **2019,** *7*, 11612–11620.

143. Gurunathan, S.; Woong Han, J.; Eppakayala, V.; Kim, J. Green Synthesis of Graphene and its Cytotoxic Effects in Human Breast Cancer Cells. *Int. J. Nanomed.* **2013,** 1015.

144. Liang, T.; Guo, X.; Wang, J.; Wei, Y.; Zhang, D.; Kong, S. Green Synthesis of Porous Graphene-Like Nanosheets for High-Sensitivity Nonenzymatic Hydrogen Peroxide Biosensor *Mater. Lett.* **2019,** *254*, 28–32.

145. Mondal, O.; Mitra, S.; Pal, M.; Datta, A.; Dhara, S.; Chakravorty, D. Reduced Graphene Oxide Synthesis by High Energy Ball Milling. *Mater. Chem. Phy.* **2015,** *161*, 123–129.

146. Ansari, M. Z.; Lone, M. N.; Sajid, S.; Siddiqui, W. A. Novel Green Synthesis of Graphene Layers using Zante Currants and Graphene Oxide. *Orient. J. Chem.* **2018,** *34*, 2832–2837.

147. Gan, L.; Li, B.; Chen, Y.; Yu, B.; Chen, Z. Green Synthesis of Reduced Graphene Oxide using Bagasse and its Application in Dye Removal: A Waste-To-Resource Supply Chain. *Chemosphere* **2019,** *219*, 148–154.

148. Wang, Y.; Shi, Z.; Yin, J. Facile Synthesis of Soluble Graphene via a Green Reduction of Graphene Oxide in Tea Solution and its Biocomposites. *ACS Appl. Mater. Interfaces* **2011,** *3*, 1127–1133.

149. Yang Zhang, H.; Shen, G.; Cheng, P.; Zhang, J.; Guo, S. Reduction of Graphene Oxide Vial-Ascorbic Acid *Chem. Commun.* **2010**, *46*, 1112–1114.
150. Bhattacharya, G.; Sas, S.; Wadhwa, S.; Mathur, A.; McLaughlin, J.; Roy, S. S. Aloe Vera Assisted Facile Green Synthesis of Reduced Graphene Oxide for Electrochemical and Dye Removal Applications *RSC Adv.* **2017**, *7*, 26680–26688.
151. Gnana kumar, G.; Justice Babu, K.; Nahm, K. S.; Hwang, Y. J. A Facile One-Pot Green Synthesis of Reduced Graphene Oxide and its Composites for Non-Enzymatic Hydrogen Peroxide Sensor Applications *RSC Adv.* **2014**, *4*, 7944.
152. Pei, S.; Wei, Q.; Huang, K.; Cheng, H. M.; Ren, W. Green Synthesis of Graphene Oxide by Seconds Timescale Water Electrolytic Oxidation *Nat. Commun.* **2018**, *9*, 1.
153. Tai, M. J. Y.; Liu, W. W.; Khe, C. S.; Hidayah, N. M. S.; Teoh, Y. P.; Voon, C. H.; Lee, H. C.; Adelyn, P. Y. P. Green Synthesis of Reduced Graphene Oxide using Green Tea Extract. *AIP Conf. Proc.* **2018**, *2045*, 020032.
154. Liu, K.; Zhang, J. J.; Cheng, F. F.; Zheng, T. T.; Wang, C.; Zhu, J. J. Green and Facile Synthesis of Highly Biocompatible Graphene Nanosheets and its Application for Cellular Imaging and Drug Delivery. *J. Mater. Chem.* **2011**, *21*, 12034.
155. Kartick, B.; Srivastava, S. K.; Srivastava, I. Green Synthesis of Graphene. *J. Nanosci. Nanotechnol.* **2013**, *13*, 4320–4324.
156. Raveendran, S.; Chauhan, N.; Nakajima, Y.; Toshiaki, H.; Kurosu, S.; Tanizawa, Y.; Tero, R.; Yoshida, Y.; Hanajiri, T.; Maekawa, T.; Ajayan, P. M.; Sandhu, A.; Kumar, D. S. Ecofriendly Route for the Synthesis of Highly Conductive Graphene using Extremophiles for Green Electronics and Bioscience. *Par. Par. Syst. Charact.* **2013**, *30*, 573–578.
157. Gurunathan, S.; Han, J. W.; Park, J. H.; Eppakayala, V.; Kim, J. H. Ginkgo Biloba: A Natural Reducing Agent for the Synthesis of Cytocompatible Graphene. *Int. J. Nanomed.* **2014**, *9*, 363–377.
158. Akhavan, O.; Ghaderi, E. Escherichia coli bacteria Reduce Graphene Oxide to Bactericidal Graphene in a Self-Limiting Manner. *Carbon* **2012**, *50*, 1853–1860.
159. Fernandez-Merino, M.; Guardia, L.; Paredes, J. Vitamin C as an Innocuous and Safe Reductant for the Preparation of Graphene Suspensions from Graphite Oxide. *J. Phys. Chem.* **2010**, *114*, 6426–6432.
160. Zhu, C.; Guo, S.; Fang, Y.; Dong, S. Reducing Sugar: New Functional Molecules for the Green Synthesis of Graphene Nanosheets. *ACS Nano.* **2010**, *4*, 2429–2437.
161. Esfandiar, A.; Akhavan, O.; Irajizad, A. Melatonin as a Powerful Bio-Antioxidant for Reduction of Graphene Oxide. *J. Mater. Chem.* **2011**, *21*, 10907–10914.
162. Wang, G.; Qian, F.; Saltikov, C. W.; Jiao, Y.; Y. Yat, Microbial Reduction of Graphene Oxide by Shewanella. *Nano. Res.* **2011**, *4*, 563–570.
163. Zhang, H.; Yu, X.; Guo, D.; Qu, B.; Zhang, M.; Li, Q.; Wang, T. Synthesis of Bacteria Promoted Reduced Graphene Oxide–Nickel Sulfide Networks for Advanced Supercapacitors. *ACS Appl. Mater. Interf.* **2013**, *5*, 7335–7340.
164. Carmalin Sophia, A.; Lima, E. C.; Allaudeen, N.; Rajan, S. Application of Graphene Based Materials for Adsorption of Pharmaceutical Traces from Water and Wastewater - a Review. *Desalination Water Treat* .**2016**, *57*, 1–14.
165. Agharkar, M.; Kochrekar, S.; Hidouri, S.; and Azeez, M. A. Trends in Green Reduction of Graphene Oxides, Issues and Challenges: A Review. *Mater. Res. Bull.* **2014**, *59*, 323–328.

166. Eda, G.; Fanchini, G.; Chhowalla, M. Large-Area Ultrathin Films of Reduced Graphene Oxid as Transparent and Flexible Electronic Material *Nat. Nanotechnol.* **2008,** *3,* 270–274.

167. Upadhyay, R. K.; Soin, N.; Bhattacharya, G.; Saha, S.; Barman, A.; and Roy, S. S. Grape Extract Assisted Green Synthesis of Reduced Graphene Oxide for Water Treatment Application *Mater. Lett.* **2015,** *160,* 355–358.

168. Kroto, H. W.; Heath, J. R.; Brien, S.C. O.; Curl, R. F.; Smalley, R. E. C_{60} Buckminsterfullerene. *Nature* **1985,** *318,* 162–163.

169. Rocha, R. C. Filho Fullerenes and their Amazing Molecular Geometry. *Química Nova na Escola.* **1996,** *4,* 7–11.

170. Yadav, B. C.; Kumar, F. R. Structure, Properties and Applications of Fullerenes. *Int. J. Nanotechnol. Appl.* **2008,** *2,* 15–24.

171. Wilson, M. A.; Pang, L. S. K.; Willet, G. D.; Fisher, K. J.; Dance, L. G. Fullerenes—preparation, properties, and carbon chemistry. *Carbon* **1992,** *30,* 675–693.

172. Pang, L. S. K.; Vassallo, A. M.; Wilson, M. A. Fullerenes from Coal. *Nature* **1991,** *352,* 480–480.

173. Pang, L. S. K.; Vassallo, A. M.; Wilson, M. A. Fullerenes from Coal: A Self-Consistent Preparation and Purification Process. *Energy Fuels* **1992,** *6,*176–179.

174. Qiu, S.; Zhou, Y.; Yang, Z. G.; Wang, D. K.; Guo, S. C.; Tsang, S. C.; Harris, P. J. F. Preparation of Fullerenes using Carbon Rods Manufactured from Chinese Hard Coals. *Fuel.* **2000,** *79,* 1303–1308.

175. Weston, A.; Murthy, M.; Lalvani, S.; Synthesis of Fullerenes from Coal. *Fuel Process. Technol.* **1995,** *45,* 203–212.

176. Weston A.; Murthy, M.; Synthesis of Fullerenes: An Effort to Optimize Process Parameters. *Carbon* **1996,** 1267–1274.

177. Mukhopadhyay, K.; Krishna, K. M.; Sharon, M.; Fullerenes from Camphor: A Natural Source. *Phys. Rev. Lett.* **1994,** *72,* 3182–3185.

178. Heymann, D. Search for Ancient Fullerenes in Anthraxolite, Shungite, and Thucolite. *Carbon* **1995,** *33,* 237–239.

179. Fang, P. H.; Wong, R. Evidence for Fullerene in a Coal of Yunnan, Southwestern China. *Mater. Res. Innov.* **1997,** *1,* 130–132.

180. Mauter, M. S.; Elimelech, M. Environmental Applications of Carbon-Based Nanomaterials. *Environ. Sci. Technol.* **2008,** *42,* 5843–5859.

181. Park, S.; Srivastava, D.; Cho, K. External Chemical Reactivity of Fullerenes and Nanotubes. *MRS Online Proc. Lib. Arch.* **200l,** *675,* 151–156.

182. Da Ros, T.; Prato, M.; Novello, F.; Maggini, M.; Banfi, E. Easy Access to Water Soluble Fullerene Derivative via 1,3-Dipolar Cycloadditions of Azomethine Ylides to C_{60}. *J. Organ. Chem.* **1996,** *61,* 9070–9072.

183. Sokolov, V. I. Chemistry of Fullerenes, Novel Allotropic Modifications of Carbon. *Russ. Chem. Bull.* **1999,** *48,* 1197–1205.

184. Kharisov, B. I.; Kharissova, O. V.; Chaves-Guerrero, L. Synthesis Techniques, Properties and Applications of Nanodiamonds. *Synth. React. Inorg. Metal Org. Nano Metal Chem.* **2010,** *40,* 84–101.

185. Chung, P. H.; Perevedentseva, E.; Cheng, C. L. The Particle Size-Dependent Photoluminescence of Nanodiamonds. *Surf. Sci.* **2007,** *601,* 3866–3870.

186. Enoki, T. Diamond-to-Graphite Conversion in Nanodiamond and Electronic Properties of Nanodiamond-Derived Carbon System. *Phys. Solid State.* **2004**, *46*, 635–640.

187. *McNaught, A. D.; Wilkinson, A.* Diamond-Like Carbon Films. In *Compendium of Chemical Terminology* (IUPAC Gold Book); *Blackwell Scientific Publications: Oxford, 1997.*

188. Huss, G. R. Meteoritic Nanodiamonds: Messengers from the Stars. *Elements* **2005**, *1*, 97–100.

189. Daulton, T. L.; Eisenhour, D. D.; Bernatowicz, T. J.; Lewis R. S.; Buseck, P. R. Genesis of Presolar Diamonds: Comparative High-Resolution Transmission Electron Microscopy Study of Meteoritic and Terrestrial Nano-Diamonds. *Geochim. Cosmochim. Acta.* **1996**, *60*, 4853–4872

190. Danilenko, V. V. On the History of the Discovery of Nanodiamond Synthesis. *Phys. Solid State* **2004**, *46*, 595–599.

191. Boudou, J. P.; Tisler, J.; Reuter, R.; Thorel, A.; Curmi, P. A. Fluorescent Nanodiamonds Derived from HPHT with a Size of Less than 10 nm. *Diam. Relat. Mater.* **2013**, *37*, 80–86.

192. Fang, X.; Mao, J.; Levin, E. M.; Schmidt-Rohr, K. Nonaromatic Core - Shell Structure of Nanodiamond from Solid-State NMR Spectroscopy. *J. Am. Chem. Soc.* **2009**, *131*, 1426–1435.

193. Schuelke, T.; Grotjohn, T. A. Diamond Polishing. *Diam. Relat. Mater.* **2013**, *32*, 17–26.

194. Zeiger, M.; Jäckel, N.; Aslan, M.; Weingarth, D.; Presser, V. Understanding Structure and Porosity of Nanodiamond-Derived Carbon Onions. *Carbon* **2015**, *84*, 584–598.

195. Alkahtani, M. H.; Alghannam, F.; Jiang, L.; Rampersaud, A. A.; Brick, R. Fluorescent Nanodiamonds for Luminescent Thermometry in the Biological Transparency Window. *Opt. Lett.* **2018**, *43*, 3317–3320.

196. Deák, P.; Aradi, B.; Kaviani, M.; Frauenheim, T.; Gali, A. Formation of NV Centers in Diamond: A Theoretical Study Based on Calculated Transitions and Migration of Nitrogen and Vacancy Related Defects. *Phys. Rev. B.* **2014**, *89*, 075203–075215.

197. Turcheniuk, K.; Mochalin, V. N. Biomedical Applications of Nanodiamond. *Nanotechnology* **2017**, *28*, 252001–2520028.

198. Meinhardt, T.; Lang, D.; Dill, H.; Krueger, A. Pushing the Functionality of Diamond Nanoparticles to New Horizons: Orthogonally Functionalized Nanodiamond using Click Chemistry. *Adv. Funct. Mater.* **2011**, *21*, 494–500.

199. Yang, G. W., Wang, J. B., Liu, Q. X. Preparation of Nano-Crystalline Diamonds using Pulsed Laser Induced Reactive Quenching. *J. Phys. Condens. Mat.* **1998**, *10*, 7923–7927.

200. Shenderova, O. A.; McGuire, G. E. Science and Engineering of Nanodiamond Particle Surfaces for Biological Applications. *Biointerphases* **2015**, *10*, 030802–030825.

201. Narayan, J.; Bhaumik, A. Novel Phase of Carbon, Ferromagnetism, and Conversion into Diamond. *J. Appl. Phys.* **2015**, *118*, 215303–2153015.

202. Xiao, J.; Liu, P.; Yang, G. W. Nanodiamonds from Coal under Ambient Conditions. *Nanoscale* **2015**, *7*, 6114–6125.

203. Das, T.; Saikia, B. K. Nanodiamonds Produced from Low-Grade Indian Coals .*ACS Sustain. Chem. Eng.* **2017**, *5*, 9619–9624.

204. Khachatryan, A. K.; Aloyan, S. G.; May, P. W.; Sargsyan, R.; Khachatryan, V. A.; Baghdasaryan, V. S. Graphite-to-Diamond Transformation Induced by Ultrasound Cavitation. *Diam. Relat. Mater.* **2008,** *17*, 931−936.

205. Wang, Y.; Hu, A. Carbon Quantum Dots: Synthesis, Properties and Applications. *J. Mater. Chem. C.* **2014,** *2*, 6921−6939.

206. Hu, S.; Tian, F.; Bai, P.; Cao, S.; Sun, J.; Yang, J. Synthesis and luminescence of Nanodiamonds from Carbon Black. *Mater. Sci. Eng. B.* **2009,** *157*, 11–14.

207. Shaw, S. J.; Schiffers, W. P.; Gentry, T. P.; Emmony, D. C. A Study of the Interaction of a Laser Generated Cavity with a Nearby Solid Boundary. *J. Phys. D: Appl. Phys.* **1999,** *32*, 612–1617.

208. Yavas, O; Schilling, A.; Bischof, J.; Boneberg, J.; Leiderer, P. Bubble Nucleation and Pressure Generation During Laser Cleaning of Surfaces. *Appl. Phys. A. Mater. Sci. Process.* **1997,** *64*, 331–339.

209. Kumar, V.; Srivastava, A. K.; Toyoda, S.; Kaur, I. Extraction of Low Toxicity Nanodiamonds from Carbonaceous Wastes Fuller. *Nanotub. Car. N.* **2015,** *24*, 190–194.

210. Schrand, A. M.; Hens, S. A. C.; Shenderova, O. A. Nanodiamond Particles: Properties and Perspectives for Bioapplications. *Crit. Rev. Solid State Mater. Sci.* **2009,** *34*, 18−74.

211. Schrand, A. M.; Huang, H.; Carlson, C.; Schlager, J. J.; Osawa, E.; Hussain, S.M.; Dai, L. Are Diamond Nanoparticles Cytotoxic? *J. Phys. Chem. B.* **2007,** *111*, 2−7.

212. Mochalin, V. N.; Shenderova, O.; Ho, D.; Gogotsi, Y. The Properties and Applications of Nanodiamonds. *Nat. Nanotechnol.* **2012,** *7*, 11−23.

213. Mochalin, V. N.; Gogotsi, Y. Nanodiamond-Polymer Composites. *Diam. Relat. Mater.* **2015,** *58*, 161−171

214. Dolmatov, V. Y. Detonation Nanodiamonds: Synthesis, Structure, Properties and Applications. *Russ. Chem. Rev.* **2007,** *76*, 339−360.

215. Shenderova, O.; Vargas, A.; Turner, S.; Ivanov, D. M.; Ivanov, M. G. Nanodiamond-Based Nanolubricants: Investigation of Friction Surfaces. *Tribol. Trans.* **2014,** *57*, 1051−1057.

CHAPTER 2

Microwave-Assisted Synthesis: A New Tool in Green Technology

SREERENJINI C. R.[1], BHAGYALAKSHMI BALAN[1], GLADIYA MANI[1], AND SURESH MATHEW[1,2*]

[1]*School of Chemical Sciences (SCS), Mahatma Gandhi University, Kottayam 696560, Kerala, India*

[2]*Advanced Molecular Materials and Research Centre (AMMRC), Mahatma Gandhi University, Kottayam 696560, Kerala, India*

Corresponding author. E-mail: sureshmathewmgu@gmail.com

ABSTRACT

Science is always advancing ever since from the earlier days and the advancement had been incremental. Chemical synthesis and manufacturing plays a vital role in the progress of science and in the earlier days, we were more focused upon yield and profit than safety and ecological effects of chemical products and processes. However later on we became aware of the detrimental impacts of harsh chemicals and chemical practices on the environment and started to develop more environment and eco-friendly synthesis methods which are direct and efficient. Microwave-assisted synthesis is a new tool in green technology in this regard. Considerably shorter reaction time, uniform product formation with high yields, superior product purity and material properties etc are the main attractive of microwave assisted synthesis methods compared to conventional chemical synthesis methods. For the past two decades, microwave assisted synthesis finds application in organic and polymer synthesis, material sciences, nanotechnology, and biochemical processes. Herein, we discuss about the history of development of microwave technology, working principles and microwave-assisted synthesis of nanomaterials for various applications like photo catalysis, propellants, and super capacitors etc.

2.1 INTRODUCTION

The emergence of nanoscience and nanotechnology along with the introduction of green chemistry has imposed new demands on material synthesis. The 12 principles of green chemistry expect the evolution of efficient synthetic routes for material synthesis with reduced reaction time, minimal use of toxic chemicals, and less waste production. In this perspective, microwave-based synthetic routes will be a step toward sustainable development. Microwave-assisted synthesis is considered as an eco-friendly faster synthesis route, due to its ability to couple the reactant molecules quickly by raising the reaction temperature. The bottleneck of conventional synthesis methods such as lengthy reaction time, inhomogeneity in shape and size, and slow reaction pace have been well addressed by microwave method.

In microwave-assisted synthesis, polar molecules in the solvent or conducting ions in a solid are forced to align or rotate with the field and to collide each other rapidly resulting in the dissipation of energy in the form of heat. This ensures uniform or homogeneous heating of the precursor material completely rather than surface to bulk heating taking place in conventional synthesis methods. The faster collision of molecules within the reaction vessel results in an accelerated reaction rate which then leads to a shorter reaction time compared to days or hours of processing time in conventional techniques. Experimental studies prove that microwave synthesis increases the reaction rate exponentially compared to conventional techniques due to the use of higher reaction temperature with lessened side reactions and byproducts formations. This results in the formation of products with high yield and purity.

In conventional techniques, the heat transfer is from the walls of the reaction to the solvent medium. This creates a temperature gradient within the sample. While in the microwave method, the radiations excite the precursors inside the reaction vessel, resulting in an even distribution of temperature within the sample rather than creating a temperature gradient. This even heating and lesser solvent requirement urge the researches to adapt to microwave-assisted synthesis. Microwave-assisted synthesis methods offer more reproducibility as they involve uniform heating and better control over reaction parameters than traditional synthesis techniques. The adoption of microwave synthesis helped to attain high yield and low processing cost. This facile method opens up chances in the evolvement of new material phases which could not be obtained by normal synthesis methods.

The interaction of materials with microwave radiations varies since there is a difference in their susceptibilities. Depending on their response, materials can be broadly classified into three categories: (i) materials that are transparent to microwave irradiation, for example, sulfur (ii) materials that absorb microwave radiation, for example, water, and (iii) materials that reflect microwave radiation, for example, metals. In microwave chemistry, the interaction of microwave radiation with the materials is a prime factor and hence microwave radiation-absorbing materials are of the utmost importance.[1] Microwave-assisted synthesis is the foremost and most suitable synthesis route for various materials like for the synthesis of nanoparticles, organic molecules, polymers, magnetic particle synthesis, and so on. Microwave synthesis methodology is widely being explored in the field of supercapacitors, photocatalysts, and propellant catalysts. Due to the availability of large microwave reactors or apparatus, it is now easy to scale up the laboratory level experiments to industrial level without altering the reaction parameters within few minutes than tedious conventional methods.

2.2   HISTORY AND DEVELOPMENT

Nowadays, microwave-assisted synthesis has received considerable attention among the scientific community. The increasing diversity and availability of microwave equipment has allowed this technology to become more popular and useful. A breakthrough in microwave technology occurred during the World War II by Dr. Percy LeBaron Spencer, who developed the first fully functioning microwave oven by joining a high-density electromagnetic field generator device to an enclosed metal box.

Later, in 1947 Raytheon developed the first commercial microwave oven "1161 Radarange" with weight nearly 750 pounds. The first kitchen counterpart, domestic oven was introduced by Amana (a division of Raytheon). It was smaller, safer, and more reliable than the previous models, costing under $500. During 1970s there was a massive increase of microwave ovens elsewhere in the world. In 1978, the first microwave laboratory instrument was developed by CEM Cooperation, USA, for analyzing moisture in solids. Up to the middle of 1980s microwave oven was only used for cooking and defrosting frozen food. Since 1983–1985 microwave radiation was used for chemical analysis.[1,2] Robert Gedye, of Laurentian University, Canada, George Majetich of University of Georgia, USA and Raymond Giguere of Mercer University, USA published papers relating to microwave synthesis.[3]

A revolutionary milestone in the history of microwave synthesis happened in 1990s, the first high pressure microwave vessel "HPV 80" was established by Milestone Srl Italy, for complete digestion of materials like oxides, oils, and pharmaceutical compounds. Later in 1992–1996, CEM Corporation developed a more efficient batch system reactor (MDS 200) and a single-mode cavity system (Star 2) for chemical synthesis. During 1997, Prof. H. M. Kingston of Duquesne University, USA culminated an innovative book titled "Microwave-Enhanced Chemistry-Fundamentals, Sample Preparation and Applications," edited by H. M. Kingston and S. J. Haswell. Since 2000 microwave chemistry emerged as a promising field of study in chemical synthesis. Companies like CEM, Biotage, Anton parr, and Milestone marketed a number of microwave reactors of varying capacities and temperature control that enlarges the applicability and prosperity of microwave-assisted synthesis.

2.3 MICROWAVE-ASSISTED SYNTHESIS OF NANOMATERIALS FOR VARIOUS APPLICATIONS

2.3.1 PHOTOCATALYSIS

Microwave synthesis has proved to be a rapid and facile method that offers effective heating source in the synthesis of nanoparticles, which gave high-quality products. Microwave not only accelerates the reaction rate but also saves energy and time.[5,6] Now, this technique has been widely used for the synthesis of metal oxide nanoparticles, which can be used in multifunctional applications including photocatalysis,[7] photoelectrochemical cells,[8] sensor devices,[9] and dye-sensitized solar cells.[10] Several metal oxides such as TiO_2,[11] ZnO,[12] MnO_2,[13] CeO_2,[14] and Fe_2O_3[15] and their composites have been extensively used as photocatalyst due to their inherent electronic structures.

With the development of photocatalysis, nanosized TiO_2 was the most used catalyst for photochemical reactions. Several synthetic strategies such as sol-gel,[16] hydrothermal synthesis,[17] and microwave methods were adopted to synthesize the TiO_2-based photocatalyst. Anirudha Jena et al. obtained a phase-pure nanocrystalline anatase TiO_2 by microwave method. They obtained unique mesoporous titania samples with spherical morphology and narrow size distribution of better catalytic activity than the commercially available TiO_2 samples.[18] A microwave-assisted synthesis of TiO_2 nanoparticles with an average size of 7 nm was reported by G. S. Falk et al. The rapid and homogenous heating of microwave radiation is capable of inducing uniform nanoparticle distribution within few minutes than conventional methods. The photocatalytic activities of the synthesized samples are comparable with TiO_2 P25.[19]

Nitrogen-doped TiO_2 is a very important visible light photocatalyst due to its stability and inexpensiveness. An in situ microwave-assisted synthesis of N-TiO_2/graphitic carbon nitride (g-C_3N_4) composite was prepared by Xiao-jing Wang. The scanning electron microscope (SEM) micrographs of the composite ensure the successful growth of N-TiO_2 on the lamellar structure of g-C_3N_4 without aggregation (Fig. 2.1a). It was again confirmed from the dark particles and grey areas of the transmission electron microscope (TEM) image. The particle with dark color can be assigned to be N-TiO_2 and the grey area was assigned to be g-C_3N_4 (Fig. 2.1b). During the calcinations the peroxotitanate releases NH_3 results in high porous structure and large surface area. The photocatalytic activities of the synthesized samples were evaluated using rhodamine B (Rh B) and methylene blue (MB), the catalytic activity was increased gradually with the content of N-TiO_2 increasing from 15 to 40 wt%. The composite with 40 wt% shows the best performance.[20]

Recently, microwave synthesis of ZnO nanoparticles was reported by the group of Chaiyos Chankaew using longan seeds biowaste. The influence of zinc precursor, particle sizes based on irradiation time, and microwave power were studied. The 80 W 30 cycles of microwave irradiation result in pure hexagonal phase of ZnO nanoparticles of 10–100 nm sizes with a specific surface area of 35 m^2/g.[21] Photocatalytic ammonia production through nitrate reduction was found to be very effective in context of current efforts. Now, Pd-doped TiO_2 nanoparticles were synthesized through microwave irradiation and they are utilized as an efficient photocatalyst for ammonia reduction. The Pd-TiO_2 with 2.65 wt% producing a high rate of ammonia, 21.2 µmol.[22] The morphologies and surface conditions of the ZnO microstructures can be controlled by microwave method. Morphology-controlled ZnO microstructures ranging from 1 to 2 µm were fabricated by varying ammonia concentration in a microwave reactor. The fast microwave-assisted synthesis of ZnO effects mesoporous structures with large specific surface area and pore volume. The SEM micrographs of the synthesized samples clearly confirmed the morphology controlled ZnO microstructures formation. The photocatalytic degradation of the samples with Rh B was increased by increasing the specific surface area.[23]

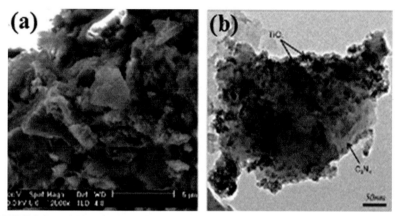

FIGURE 2.1   (a) SEM images of the 40 wt% N-TiO$_2$/g-C$_3$N$_4$ composite, (b) TEM images of the 40 wt% N-TiO$_2$/g-C$_3$N$_4$ composite.
Source: Reproduced with permission from Ref. [20]. Copyright 2013 American Chemical Society.

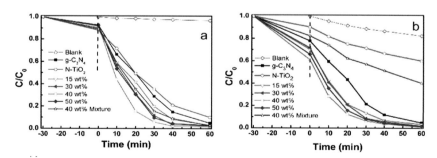

FIGURE 2.2   (a) Photocatalytic activities of N-TiO$_2$, g-C$_3$N$_4$, and N-TiO$_2$/g-C$_3$N$_4$ composites on the degradation of (a) Rh B and (b) MB under visible light irradiation.
Source: Reproduced with permission from Ref. [20]. Copyright 2013 American Chemical Society.

As a typical narrow band gap semiconductor photocatalysts, tungsten oxides (WO_x, 2.4–2.8 eV) are very attractive than other metal oxides.[24,25] The microwave-assisted solvothermal synthesis of stacked orthorhombic $WO_3.H_2O$ and urchin-like monoclinic $W_{18}O_{49}$ nanowires were reported by Arpan Kumar Nayak et al. Both the photocatalytic degradation and hydrogen evolution activity of the synthesized samples was evaluated in neutral medium. The large surface area, oxygen vacancies, and fast charge transport properties enhance their catalytic activity.[26]

Ferric oxides, mainly α-Fe_2O_3 display significant interest in photocatalytic applications. Their morphology and particle size play an important role in determining the activity of α-Fe_2O_3. A nanosized α-Fe_2O_3 powder was synthesized by microwave-assisted hydrothermal reaction. This approach contributes new synthetic approach for the synthesis of nanosized α-Fe_2O_3 and uniform particles of about 5 nm in size were formed with a surface area of about 173.0 m^2g^{-1}. The catalyst exhibited excellent catalytic performance for the oxidation of CO and 2-propanol to CO_2.[27] A simple one-step NaCl-assisted microwave solvothermal method was adopted for the synthesis of α-Fe_2O_3 monodisperse microspheres. The advantages of this microwave method were to control the size of microspheres by changing the microwave-solvothermal time. The high resolution transmission electron microscope (HR-TEM) micrographs of the synthesized samples show that the average size of the sample was found to be smaller than 5 nm.[28]

Like iron oxides, bismuth oxides can also be a promising candidate for the degradation of organic species under visible light. Juliana S. Souza et al. reported Au-doped bismuth vanadate nanoflowers through conventional and microwave methods. It was found to be the composites exhibited same physical–chemical properties as those prepared through conventional heating. They establish how microwaves can replace the well-established methods to synthesize inorganic nanomaterials and reduce both energy and time. The composites show excellent catalytic activity to degrade 95% of MB under UV–visible light irradiation.[29]

Currently, metal-organic framework derived (MOF) porous metal oxides are attractive in the field of sensors, catalysis, electrochemical, and photochemical devices. A simple and fast microwave-assisted synthesis of reduced graphene oxide incorporated MOF derived ZnO was reported by GuangZh et al. The photocatalytic degradation of MB using this composite was achieved with 82% efficiency. The enhanced performance of the composite was attributed to high light absorption and the separation of photogenerated electron-hole pairs. It was clearly observed in the photoluminescence (PL) spectrum.[30]

2.3.2   PROPELLANTS

Ammonium perchlorate (AP) is the most common oxidizer used in composite solid propellants (CSPs) and it comprises almost 60–70% of the total mass of the propellant. Being the major component of propellant fuel, the thermal behavior pattern of AP directly influences the propellant performance. Burn rate modifiers are the commonly used additives in propellants to accelerate the propellant burn rate and AP is exceptionally sensitive to the presence of these additives. Burn rate modifiers can tailor or alter the ballistic properties of the propellant and thus its performance. Transition metal oxides (TMOs) such as copper oxide (CuO), copper chromite ($CuCr_2O_4$), and iron oxide are the conventionally used burn rate modifiers.

It is known that the synthetic routes can improve or tune the properties of materials especially their crystalline nature, uniform shape and size, morphology, surface area, and electronic properties. These tuning of properties makes them a prospective candidate for versatile applications. For example, rod-shaped CuO exhibited enhanced specific capacitance than CuO with irregular shape due to increased surface area.[31] Hence much effort has been devoted by the researchers to the synthesis of metal oxide nanomaterials with controlled physical properties. Yang et al. reported that the microwave-assisted synthesis of CuO nanoparticles could impart significant improvement in the crystallinity and crystallite size of CuO than the one prepared via chemical precipitation.

Compared to the shortcomings suffer from conventional synthesis methods, microwave synthesis endows facile synthesis route, reduced reaction time, homogeneous heating, energy-saving, and uniform size distribution. It was found that after microwave hydrothermal modification the morphology of synthesized CuO got tuned to rod structure which could be attributed to Ostwald ripening. Being a good microwave absorbent, CuO speeds up the dissolution of smaller CuO crystallites and thus the growth of larger ones. Later, the catalytic efficiencies of microwave modified samples were studied via the thermal decomposition of AP and it was found that by the addition of modified CuO the thermal decomposition temperature of AP was reduced by almost 100°C compared to the unmodified CuO.[32]

By combining the microwave method with the hydrothermal approach, we can achieve an amalgamation of uniform and fast heating of microwave method with hydrothermal crystallization allows the formation of nanoparticles with uniform size and morphology.[33] For example, via microwave hydrothermal approach urchin-like CuO was synthesized within a short span of 15 min[34] and there are so many successful cases where microwave-assisted synthesis approach resulted in the formation of CuO nanoparticles with exceptional properties which makes this method a novel one.

Su and group reported that hyphenation of microwave and hydrothermal routes could result in the formation of CuO nanosheets and dendrites. They found that hierarchical dendrite-like $Cu(OH)_2$ nanostructure was formed after hydrolysis of $CuSO_4$ in the presence of PEG(Polyethelene Glycol)-400. After microwave hydrothermal treatment, the dendrite morphology was disassembled to discrete CuO nanosheets. Here the role of microwave irradiation is to destroy the integrating force between nanosheets aggregates which resulted in the disassembly of the assembled structure. Later, the synthesized CuO nanostructures were found to be a promising catalyst for the decomposition studies of AP due to their large surface-to-volume ratio. It was observed that the synthesized CuO nanostructures could effectively reduce the decomposition temperature of AP to around 350°C from 439°C of pure AP.[33]

Cu_2O is another renowned TMO with different structures and was prepared via the microwave-assisted solvothermal method by Luo et al. simply by adjusting the composition of solvents. Compared to the conventional time-consuming synthesis routes, microwave irradiation makes the synthesis procedure a facile one. This makes the metal oxide synthesis an effortless one and thus opens up new arenas in the field of material synthesis. This synthesis route resulted in three types of Cu_2O cubes: cubic aggregates, monodispersed cubes, and {100} plane etched cubes which could successfully change the activation energy, E during the course of the decomposition process. The monodispersed Cu_2O cubes exhibited efficient catalytic activity in triggering the AP decomposition while {100} plane etched cubes exhibited the least energy of activation during the decomposition process.[35]

Gong et al. also prepared Cu_2O for thermal decomposition studies of AP and by using microwave method they could complete the tedious synthesis process in less than 45 min. Implementation of the microwave method saves a lot of preparation time of researchers and makes it a superficial or effortless one. Similar to all conventional methods, microwave method also offers lots of opportunities in the optimization of reaction conditions like time of reaction, solvents used, power applied during the reaction procedure, etc. Each change in the above factors makes considerable difference in the properties of the synthesized materials and thus microwave method the area of synthetic research and the experimental results indicate that Cu_2O has better catalytic activity for AP decomposition.[36]

Apart from CuO and Cu_2O, metallic Cu is also finding tremendous applications in modern electronic circuits and catalysis due to excellent properties. Even though the thermal decomposition behavior of AP can be significantly improved to an extent by particle size reduction of AP, it is a dangerous process. Hence nanosized metal oxides or metal particles have proved effective to catalyze the thermal decomposition of AP. In contrast, the pure metal nanoparticle is more active in the thermal decomposition of AP than metal-oxide nanoparticles. With the aid of microwave irradiation, metallic Cu nanoparticles were prepared using a time period of less than 5 min. Experimental observation indicates that synthesized Cu nanoparticles are better catalyst than the conventional catalyst. Thermal studies using thermo gravimetric analysis (TG) and differential scanning calorimetry (DSC) reveal that nanoscale Cu particles effectively lowers the energy of activation for AP, composite solid propellants, high melting explosives (HMX) and nitrotriazolone (NTO). The burning rate of CSPs was also increased by using Cu particles.[37]

Graphene is an atom-thick flat sheet which is considered as the mother form or building unit of all the graphitic forms like 0D buckyballs, 1D nanotubes, and 3D graphite. Hence very recently graphene or graphene sheet based composite is catching the attention of the research community. Graphene sheets are a wonderful substrate for the dispersion and stabilization of synthesized nanoparticles without agglomeration. Microwave irradiation is considered as the most environmentally benign technique for this graphene-based composite preparation. Graphene–iron oxide nanocomposite (GINC) was successfully synthesized via microwave synthesis. Morphology studies revealed that the synthesized Fe_2O_3 particles are in the nanoregime and the crystalline nature was confirmed by selected area electron diffraction (SAED) pattern. By the use of GINC in the propellant formulation, the burn rate value was significantly increased (52%) compared to microsized Fe_2O_3 (30%) and nanosized Fe_2O_3 (37%) and found to be an excellent burn rate modifier for advanced AP based propellant system.[38] Similarly, Abhiit Day and his coworkers developed graphene-TiO_2 nanocomposite (GTNC) for propellants via microwave-assisted green process. Later, the thermal analysis identified that GTNC is an effective burn rate enhancer for AP-based solid propellants.[39] These studies clearly depict that even simple synthesis procedures can lead to the development of novel efficient catalysts with versatile applications. Sometimes the simple synthetic routes give rise to better results than tedious synthetic procedures. Microwave-based green procedures help the research community in the development of novel materials for novel applications for the development of sustainable society.

The metal oxide or metal decorated carbon nanotubes (CNTs) are also known for their excellent catalytic activity. CNTs have decent microwave properties due to their lightweight, high thermal stability, etc. Encapsulation or attachment of metal nanoparticles on CNT enhances their properties.[40] Homogeneously dispersed Co/CNT nanocomposite was prepared by Zhang et al. using a microwave-assisted method. The aim of this work was to examine the catalytic activity of the prepared composite for the thermal decomposition of AP. The experimental results say that the Co/CNT composite prepared via microwave irradiation method was a better catalyst than that of pure Co nanoparticle and Co/CNT composite prepared via the water bath method. The surface area analysis says that compared to water bath prepared composite, microwave irradiation led to higher surface material which was favorable for its higher catalytic effect.[41] Similarly, Ni/CNT prepared by microwave irradiation had a compact coating, high nickel loading, and large surface area which made it a potential burn rate modifier for AP-based propellants.[42]

Synthesis of polyhedron ZnO microparticles was achieved using microwave irradiation. The microwave method assisted in the arrangement of hexagonal unit cells in a polygonal format which later formed a polyhedral structure ZnO. This specific morphology of ZnO made it a prospective catalyst for thermal decomposition studies of AP and catalytic reduction of 2-nitrophenol and 4-nitrophenol. The study reveals that ZnO particles possess high catalytic activity since a catalyst loading of only 1% could effectively reduce the thermal degradation temperature by 30°C.[43]

Likewise, CeO_2 octahedra were also synthesized by a facile microwave mediated hydrothermal method. This microwave-assisted hydrothermal method can increase the effective collision rate and thus reducing the time required for nuclei growth because fast microwave heating can enhance the kinetics of nucleation and facet growth rate.

Even though there are many reports on CeO_2 octahedra synthesis, all of them are nonuniform in size and distribution due to practice agglomeration. Even worse, they all took on average of 6–24 synthesis hours for successfully completing the reaction procedure. Upon addition of 4 wt% of microwave synthesized CeO_2 octahedra, the decomposition temperature was reduced by 52°C which implies the high catalytic activity of synthesized CeO_2 octahedra.[44]

Very recently g-C_3N_4 is receiving ample attention from the research world due to their inherent features which makes it a potential candidate in the catalysis world. High surface area is one of the most important prerequisites for material to act as a catalyst, for the reason that as the surface area increases, catalytic efficiency also increases. Mesoporous-g-C_3N_4 is an astound material with an exceptional surface area. Usually, m-g-C_3N_4 has obtained by template supported synthesis methods especially hard-template assisted synthesis methods that involve the use of harmful chemicals like hydrogen fluoride (HF).

Zhang et al. have reported the synthesis of m-g-C_3N_4/CuO through a microwave pretreatment and mixing-calcination method. Compared to pure g-C_3N_4, microwave-assisted g-C_3N_4, CuO and g-C_3N_4/CuO, and the m-g-C_3N_4/CuO nanocomposite effectively triggered the thermal decomposition of AP. By the addition of m-g-C_3N_4/CuO nanocomposite, the decomposition temperature was lowered considerably and also the heat-released got increased exponentially, that is, from 310.5 J/g (for pure AP) to 2116.3 J/g (for the composite). The catalytic mechanism suggests that the enhancement of catalytic activity was solely due to the microwave-assisted synthesis and synergy between m-g-C_3N_4 and CuO.[45]

2.3.3 SUPERCAPACITORS

Increasing demand of energy is one of the crucial challenges faced by humankind subsequent to the industrial revolution. Development of supercapacitors was one of such endeavors to tackle this and found to be very advantageous. For a material to exhibit superior super capacitance, refined and controlled porosity, nanostructure, and high surface area are a must. These fundamental parameters are mainly depending upon the synthesis strategy adopted for the development of parent material. Conventional methods which are in operation for the synthesis of metal oxide nanoparticles for supercapacitors are: wet chemical synthesis, precipitation method, hydrothermal method, etc. Though these methods are functioning, profound time requirement, wastage of chemicals, and uncontrollability over the nanostructuring of materials are major difficulties faced by the researchers.

The emergence of green chemistry principles put forward microwave-assisted synthesis of metal oxide nanoparticles as a solution to address this scenario. After the launching of this advanced and clean chemical synthesis route, an ever-growing number of studies have been released regarding the microwave-assisted synthesis of metal oxide nanoparticles for several applications especially for supercapacitor applications. This fast, efficient, and environment-friendly synthetic strategy employs microwave irradiation as a nonconventional heating source and also, it is a noncontacting method.

Though there are several metal oxides, transition metal oxides are widely being explored for supercapacitor applications since they are capable of exhibiting multiple oxidation states and facilitate numerous redox reactions. Co_3O_4 is a renowned supercapacitor electrode material with multiple nanostructures like rods, wires, tubes, and sheets depending upon their synthesis route. High specific capacitance and longer cyclic lifetime make it a superior supercapacitor electrode material over the expensive competitor RuO_2. Jun Yan et al.[46] synthesized a graphene nanosheet/Co_3O_4 composite for supercapacitors via microwave-assisted synthesis method. The microwave-assisted synthesis route helped them to achieve well-dispersed Co_3O_4 particles (3–5 nm) on graphene nanosheet. This greatly enhanced the electrochemical and charge–discharge performance of the electrode maintaining electrical conductivity. The electrodes exhibited maximum specific capacitance of 243.2 F/g at a scan rate of 10 mV/s in 6 M aqueous potassium hydroxide solutions with approximately 95.6% specific capacitance retention even after 2000 cyclic tests.

Nanostructured TiO_2-activated carbon composite electrodes for supercapacitors were synthesized by M. Selvakumara and D. Krishna Bhat[47] and they have employed microwave irradiation to convert the precursors to TiO_2 nanoparticles. During the process of synthesis, they have found that the microwave irradiation ensures the rapid mixing and superheating effectively and thereby confirmed the uniform product formation within no time compared to the other conventional method of synthesis. The prepared nanocomposites electrodes exhibited good power and capacitive behavior and a specific capacitance of 122 F/g with high cyclic stability.

Microwave irradiation is capable of increasing the temperature of materials to a higher temperature range in a short time span, providing energy to the solution and hence it is advantageous with cost effective for bulk synthesis.[48–51] Composite made of NiO and MnO_2 nanoparticles decorated on the surface of graphite has been synthesized by Yuhong Bi et al.[52] via microwave-assisted synthesis route and obtained a higher specific capacitance of 102.78 mAh/g at 1 A/g with superior cycling stability and specific capacitance retention of 140% over 1500 cycles. The practice of microwave irradiation helped them to achieve uniform particle distribution with high purity and better reproducibility.

Thibeorchews Prasankumar et al.[53] synthesized an Mn/Zn bimetallic oxide for the very first time via microwave-assisted synthesis route for supercapacitor applications. The adoption of this kind of green and eco-friendly technique helped them to accomplish high yield and scalability. Through their work, they can improve the electrochemical performance of MnO_2 by the incorporation of conducting ZnO three dimensional frameworks. They could successfully fabricate an electrode with a high specific capacitance of 250 F/g at 0.5 A/g and long-term cycling life of 5000 cycles in 1 M KOH solution. Likewise, Kalimuthu Vijaya Sankar also synthesized Mn_3O_4 nanoparticles by microwave-assisted synthesis route within a very short reaction time of 5 min for the very first time.[54] They have also checked the supercapacitor performance of electrode in various electrolytes and found that the better performance is offered in 6 M KOH solutions with a specific capacitance of 94 F/g.

A combination of facile solvothermal and microwave-assisted synthesis route was employed by Weize Yang et al.[55] for the fabrication of a 3D Co_2 $(OH)_3Cl-MnO_2$ hybrid spheres. This brisk method facilitated the formulation of 3D sphere structures with a flowerlike morphology concocted by nanosheets as shown in Figure 2.3. On the basis of this structure, they had also put forward a possible mechanism (Fig. 2.4, Left). As a result of this, accelerated transportation of charge and ions was accomplished and ultimately it resulted in a high capacitance of 942.2 F/g. The asymmetric supercapacitor device fabricated by the synthesized hybrid material and carbon electrodes yielded a specific capacitance of 49.26 F/g and a high energy density of 12.62 Wh/kg at a power density of 679.10 W/kg. The hybrid material also exhibited improved cyclic stability of capacitance retention of 83.77% even after 5000 cycles as shown in Figure 2.4.

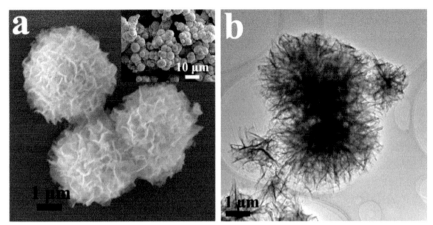

FIGURE 2.3 (a) SEM and (b) TEM images of the $Co_2(OH)_3Cl$-MnO_2 hybrid material.
Source: Reproduced with permission from Ref. [55]. Copyright 2014 American Chemical Society.

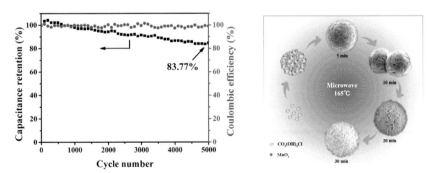

FIGURE 2.4 Cycling performance of the asymmetric supercapacitor at a current density of 2 A/g (Left) and possible formation mechanism of the $Co_2(OH)_3Cl$-MnO_2 hybrid material (Right).
Source: Reproduced with permission from Ref. [55]. Copyright 2014 American Chemical Society.

Thus in this section, we have discussed about the importance of microwave irradiation on the synthesis of metal oxide nanoparticles, specifically for supercapacitor applications and we have found that adoption of this clean, green, and environment-friendly microwave synthesis technique has a great effect on the morphology and nanostructuring of the particles and thereby have a great capability to enhance its super capacitance behavior successfully with less reaction time and high reproducibility.

2.4   CONCLUSION AND FUTURE PERSPECTIVE

The naissance and progressive trail of microwave-assisted synthesis of metal oxide nanoparticles, a clean and green technology tool have been highlighted in this chapter. This easy, facile, and fast environment-friendly method ensures clean and controllable material production at the minimal expense of time and chemicals. Microwave radiations act as a nonconventional heating source and several metal oxide nanoparticles and hybrid materials have been developed for photocatalysis, propellants, and supercapacitor applications. The adoption of microwave synthesis route is capable of fabricating fine quality products with significantly improved uniform and fine nanostructures with precise control over chemical composition surface area and interfacial characteristics. In metal oxide nanoparticle synthesis, controlling the nanostructuring of the metal oxide nanoparticle is indeed of great importance and microwave-assisted synthesis of metal oxide nanoparticles transpires as a productive and effective tool in line with green technology and environment safety.

KEYWORDS

- **microwave-assisted synthesis**
- **green technology**
- **photocatalysis**
- **propellants**
- **supercapacitors**
- **metal oxide nanoparticles**

REFERENCES

1. Grewal, A. S.; Kumar, K.; Redhu, S.; Bhardwaj, S. Microwave-Assisted Synthesis: A Green Chemistry Approach. *Int. Res. J. Pharm. App Sci*. **2013,** *3*, 278–285.
2. Gedye, R.; Smith, F.; Westaway, K.; Ali, H.; Baldisera, L.; Laberge, L.; Rousell, J. The Use of Microwave Ovens for Rapid Organic Synthesis. *Tetrahedron Lett*. **1987,** *27* (3), 279–282.
3. Gedye, R.; Smith, F.; Westaway, K.; Ali, H.; Baldisera, L.; Laberge, L.; Rousell, J. The Use of Microwave Ovens for Rapid Organic Synthesis. *Tetrahedron Lett*. **1986,** *27* (3), 279–282.
4. Horeis, G.; Pichler, S.; Stadler, A.; Gössler, W.; Kappe, C. O *Microwave-Assisted Organic Synthesis—Back to the Roots*. In 5th International Electronic Conference on Synthetic Organic Chemistry (ECSOC-5), 2001, http://www.mdpi.net/ecsoc/ecsoc-5/Papers/e0000/e0000.htm.
5. Polshettiwar, V.; Nadagouda, M. N.; Varma, R. S. Microwave-Assisted Chemistry: A Rapid and Sustainable Route to Synthesis of Organics and Nanomaterials. *Aust. J. Chem.* **2009,** *62* (1), 16–26.
6. Lagashetty, A.; Havanoor, V.; Basavaraja, S.; Balaji, S. D.; Venkataraman, A. Microwave-Assisted Route for Synthesis of Nanosized Metal Oxides. *Sci. Technol. Adv. Mater.* **2007,** *8* (6), 484.
7. Pan, L.; Liu, X.; Sun, Z.; Sun, C. Q. Nanophotocatalysts via Microwave-Assisted Solution-Phase Synthesis for Efficient Photocatalysis. *J. Mater. Chem. A* **2013,** *1* (29), 8299–8326.
8. Qiu, Y.; Yan, K.; Deng, H.; Yang, S. Secondary Branching and Nitrogen Doping of ZnO Nanotetrapods: Building a Highly Active Network for Photoelectrochemical Water Splitting. *Nano Lett.* **2011,** *12* (1), 407–413.
9. Mirzaei, A.; Neri, G. Microwave-Assisted Synthesis of Metal Oxide Nanostructures for Gas Sensing Application: A Review. *Sens. Actuators B Chem.* **2016,** *237*, 749–775.
10. Wang, H. E.; Zheng, L. X.; Liu, C. P.; Liu, Y. K.; Luan, C. Y.; Cheng, H.; Bello, I. Rapid Microwave Synthesis of Porous TiO_2 Spheres and Their Applications in Dye-Sensitized Solar Cells. *J. Phys. Chem. C*. **2011,** *115* (21), 10419–10425.
11. Gupta, S. M.; Tripathi, M. A Review of TiO_2 Nanoparticles. *Sci. Bull.* **2011,** *56* (16), 1639.
12. Ong, C. B.; Ng, L. Y.; Mohammad, A. W. A Review of ZnO Nanoparticles as Solar Photocatalysts: Synthesis, Mechanisms and Applications. *Renew. Sustain. Energy Rev.* **2018,** *81*, 536–551
13. Hoseinpour, V.; Souri, M.; Ghaemi, N. Green Synthesis, Characterisation, and Photocatalytic Activity of Manganese Dioxide Nanoparticles. *Micro Nano Lett.* **2018,** *13* (11), 1560–1563.
14. Balavi, H.; Samadanian-Isfahani, S.; Mehrabani-Zeinabad, M.; Edrissi, M. Preparation and Optimization of CeO_2 Nanoparticles and its Application in Photocatalytic Degradation of Reactive Orange 16 Dye. *Powder Technol.* **2013,** *249*, 549–555.
15. Xu, P.; Zeng, G. M.; Huang, D. L.; Feng, C. L.; Hu, S.; Zhao, M. H.; Liu, Z. F. Use of Iron Oxide Nanomaterials in Wastewater Treatment: A Review. *Sci. Total Environ.* **2012,** *424*, 1–10.

16. Behnajady, M. A.; Eskandarloo, H.; Modirshahla, N.; Shokri, M. Investigation of the Effect of Sol–Gel Synthesis Variables on Structural and Photocatalytic Properties of TiO$_2$ Nanoparticles. *Desalination*, **2011,** *278* (1–3), 10–17.

17. Hidalgo, M. C.; Aguilar, M.; Maicu, M.; Navío, J. A.; Colón, G. Hydrothermal Preparation of Highly Photoactive TiO$_2$ Nanoparticles. *Catal. Today* **2007,** *129* (1–2), 50–58.

18. Jena, A.; Vinu, R.; Shivashankar, S. A.; Madras, G. Microwave Assisted Synthesis of Nanostructured Titanium Dioxide with High Photocatalytic Activity. *Ind. Eng. Chem. Res.* **2010,** *49* (20), 9636–9643.

19. Falk, G. S.; Borlaf, M.; López-Muñoz, M. J.; Fariñas, J. C.; Neto, J. R.; Moreno, R.; Microwave-Assisted Synthesis of TiO$_2$ Nanoparticles: Photocatalytic Activity of Powders and Thin Films. *J. Nanoparticle Res.* **2018,** *20* (2), 23.

20. Wang, X. J.; Yang, W. Y.; Li, F. T.; Xue, Y. B.; Liu, R. H.; Hao, Y. J. In Situ Microwave-Assisted Synthesis of Porous N-TiO$_2$/g-C$_3$N$_4$ Heterojunctions with Enhanced Visible-Light Photocatalytic Properties. *Ind. Eng. Chem. Res.* **2013,** *52* (48), 17140–17150.

21. Chankaew, C.; Tapala, W.; Grudpan, K.; Rujiwatra, A. Microwave Synthesis of ZnO Nanoparticles Using Longan Seeds Biowaste and Their Efficiencies in Photocatalytic Decolorization of Organic Dyes. *Environ. Sci. Pollut. Res. Int.* **2019,** *26* (17), 1–7.

22. Walls, J. M.; Sagu, J. S.; Wijayantha, K. U. Microwave Synthesised Pd–TiO$_2$ for Photocatalytic Ammonia Production. *RSC Adv.* **2019,** *9* (11), 6387–6394.

23. Kim, H. B.; Jeong, D. W.; Jang, D. J. Morphology-Tunable Synthesis of ZnO Microstructures Under Microwave Irradiation: Formation Mechanisms and Photocatalytic Activity. *CrystEngComm* **2016,** *18* (6), 898–906.

24. Li, W.; Da, P.; Zhang, Y.; Wang, Y.; Lin, X.; Gong, X.; Zheng, G. WO$_3$ Nanoflakes for Enhanced Photoelectrochemical Conversion. *ACS Nano* **2014,** *8* (11), 11770–11777

25. Kalanur, S. S.; Hwang, Y. J.; Chae, S. Y.; Joo, O. S. Facile Growth of Aligned WO$_3$ Nanorods on FTO Substrate for Enhanced Photoanodic Water Oxidation Activity. *J. Mater. Chem. A* **2013,** *1* (10), 3479–3488.

26. Nayak, A. K.; Pradhan, D.; Microwave-Assisted Greener Synthesis of Defect-Rich Tungsten Oxide Nanowires with Enhanced Photocatalytic and Photoelectrochemical Performance. *J. Phys. Chem. C* **2018,** *122* (6), 3183–3193.

27. Qiu, G.; Huang, H.: Genuino, H.; Opembe, N.; Stafford, L.; Dharmarathna, S.; Suib, S. L. Microwave-Assisted Hydrothermal Synthesis of Nanosized α-Fe$_2$O$_3$ for Catalysts and Adsorbents. *J. Phys. Chem. C* **2011,** *115* (40), 19626–19631.

28. Cao, S. W.; Zhu, Y. J. Monodisperse α-Fe$_2$O$_3$ Mesoporous Microspheres: One-step NaCl-Assisted Microwave-Solvothermal Preparation, Size Control and Photocatalytic Property. *Nanoscale Res. Lett.* **2011,** *6* (1), 1.

29. Souza, J. S.; Hirata, F. T.; Corio, P. Microwave-Assisted Synthesis of Bismuth Vanadate Nanoflowers Decorated with Gold Nanoparticles with Enhanced Photocatalytic Activity. *J. Nanoparticle Res.* **2019,** *21* (2), 35.

30. Zhu, G.; Li, X.; Wang, H.; Zhang, L. Microwave Assisted Synthesis of Reduced Graphene Oxide Incorporated MOF-Derived ZnO Composites for Photocatalytic Application. *Catal. Commun.* **2017,** *88,* 5–8.

31. Gou, X.; Wang, G.; Yang, J.; Park, J.; Wexler, D. Chemical Synthesis, Characterisation and Gas Sensing Performance of Copper Oxide Nanoribbons. *J. Mater. Chem.* **2008,** *18,* 965–969.

32. Yang, C.; Xiao, F.; Wang, J.; Su, X. Synthesis and Microwave Modification of CuO Nanoparticles: Crystallinity and Morphological Variations, Catalysis, and Gas Sensing. *J. Colloid Interface Sci*. **2014**, *435*, 34–42.

33. Yang, C.; Wang, J.; Xiao, F.; Su, X. Microwave Hydrothermal Disassembly for Evolution from CuO Dendrites to Nanosheets and Their Applications in Catalysis and Photocatalysis. *Powder Technol*. **2014**, *264*, 36–42.

34. Volanti, D. P.; Orlandi, M. O.; Andrés, J.; Longo, E. Efficient Microwave-Assisted Hydrothermal Synthesis of CuO Sea Urchin-like Architectures *via* a Mesoscale Self-Assembly. *CrystEngComm* **2010**, *12*, 1696–1699.

35. Luo, X. L.; Wang, M. J.; Yun, L.; Yang, J.; Chen, Y. S. Structure-Dependent Activities of Cu_2O Cubes in Thermal Decomposition of Ammonium Perchlorate. *J. Phys. Chem. Solids*. **2016**, *90*, 1–6.

36. Gong, X. Y.; Gao, Y. P.; Wang, L. B.; Guo, P. F. Microwave-Assisted Synthesis and Catalytic Performance of Hierarchical Cu_2O Nanostructures. *Nano Brief Rep. Rev.* **2013**, *8*, 1350047.

37. Dubey, R.; Srivastava, P.; Kapoor, I. P. S.; Singh, G. Synthesis, Characterization and Catalytic Behavior of Cu Nanoparticles on the Thermal Decomposition of AP, HMX, NTO and Composite Solid Propellants, Part 83. *Thermochim. Acta* **2012**, *549*, 102–109.

38. Dey, A.; Athar, J.; Varma, P.; Prasant, H.; Sikder, A. K.; Chattopadhyay, S. Graphene-Iron Oxide Nanocomposite (GINC): An Efficient Catalyst for Ammonium Perchlorate (AP) Decomposition and Burn Rate Enhancer for AP Based Composite Propellant. *RSC Adv.* **2015,** *5*, 1950–1960.

39. Dey, A.; Nangare, V.; More, P. V.; Khan, M. A. S.; Khanna, P. K.; Sikder, A. K.; Chattopadhyay, S. A Graphene Titanium Dioxide Nanocomposite (GTNC): One-Pot Green Synthesis and Its Application in a Solid Rocket Propellant. *RSC Adv*. **2015**, *5*, 63777–63785.

40. Sui, J.; Zhang, C.; Li, J.; Yu, Z.; Cai, W. Microwave Absorption and Catalytic Activity of Carbon Nanotubes Decorated with Cobalt Nanoparticles. *Mater. Lett.* **2012**, *75*, 158–160.

41. Zhang, X.; Jiang, W.; Song, D.; Liu, Y.; Geng, J.; Li, F. Preparation and Catalytic Activity of Co/CNTs Nanocomposites *via* Microwave Irradiation. *Propell. Explos. Pyrotech.* **2009**, *34*, 151–154.

42. Zhang, X.; Jiang, W.; Song, D.; Liu, J.; Li, F. Preparation and Catalytic Activity of Ni/CNTs Nanocomposites Using Microwave Irradiation Heating Method. *Mater. Lett.* **2008**, *62*, 2343–2346.

43. Jamil, S.; Janjua, M. R. S. A.; Khan, S. R.; Jahan, N. Synthesis, Characterization and Catalytic Application of Polyhedron Zinc Oxide Microparticles. *Mater. Res. Express* **2017**, *4*, 15902–15911.

44. Shi, J.; Wang, H.; Liu, Y.; Ren, X.; Sun, H.; Lv, B. Rapid Microwave-Assisted Hydrothermal Synthesis of CeO_2 Octahedra with Mixed-Valence States and Their Catalytic Activity for Thermal Decomposition of Ammonium Perchlorate. *Inorg. Chem. Front.* **2019**, *6*, 1735–1743.

45. Zhang, Y.; Li, K.; Liao, J.; Wei, X.; Zhang, L. Microwave-Assisted Synthesis of Graphitic Carbon Nitride/CuO Nanocomposites and the Enhancement of Catalytic

Activities in the Thermal Decomposition of Ammonium Perchlorate. *Appl. Surf. Sci.* **2020**, *499*, 143875–143884.

46. Yan, J.; Wei, T.; Qiao, W.; Shao, B.; Zhao, Q.; Zhang, L.; Fan, Z. Rapid Microwave-Assisted Synthesis of Graphene Nanosheet/Co_3O_4 Composite for Supercapacitors. *Electrochim. Acta* **2010**, *55* (23), 6973–6978.

47. Selvakumar, M.; Bhat, D. K. Microwave Synthesized Nanostructured TiO_2-Activated Carbon Composite Electrodes for Supercapacitor. *Appl. Surf. Sci.* **2012**, *263*, 236–241.

48. Liu, X.; Pan, L.; Lv, T.; Zhu, G.; Lu, T.; Sun,Z.; Sun, C. Microwave-Assisted Synthesis of TiO2-Reduced Graphene Oxide Composites for the Photocatalytic Reduction of Cr(VI). *RSC Adv.* **2011**, *1*, 1245–1249.

49. Murugan, A. V.; Muraliganth, T.; Manthiram, A. Rapid, Facile Microwave-Solvo Term Synthesis of Graphene Nanosheets and Their Polyaniline Nanocomposites for Energy Storage. *Chem. Mater.* **2009**, *21*, 5004–5006.

50. Jiang, F. Y.; C. M. Wang, C. M.; Y. Fu, Y.; Liu, R. C. Synthesis of Iron Oxide Nanocubes via Microwave-Assisted Solvothermal Method. *J. Alloy Compd.* **2010**, *503*, L31–L33.

51. Liu, X.; Pan, L.; Lv, T.; Zhu, G.; Sun, Z.; Sun, C. Microwave-Assisted Synthesis of ZnO Graphene Composite for Photocatalytic Reduction of Cr(VI). *Catal. Sci. Technol.* **2011**, *1* (07),1189–1193.

52. Bi, Y.; Nautiyal, A.; Zhang, H.; Luo, J.; Zhang, X. One-Pot Microwave Synthesis of NiO/MnO2 Composite as a High-Performance Electrode Material for Supercapacitors. *Electrochim. Acta* **2018**, *260*, 952–958.

53. Prasankumar, T.; Aazem, V. I.; Raghavan, P.; Ananth, K. P.; Biradar, S.; Ilangovan, R.; Jose, S. Microwave Assisted Synthesis of 3D Network of Mn/Zn Bimetallic Oxide-High Performance Electrodes for Supercapacitors. *J. Alloys Compd.* **2017**, *695*, 2835–2843.

54. Sankar, K. V.; Kalpana, D.; Selvan, R. K. Electrochemical Properties of Microwave-Assisted Reflux-Synthesized Mn_3O_4 Nanoparticles in Different Electrolytes for Supercapacitor Applications. *J. Appl. Electrochem.* **2012**, *42* (7), 463–470.

55. Lei, Y.; Li, J.; Wang, Y.; Gu, L.; Chang, Y.; Yuan, H; Xiao, D. Rapid Microwave-Assisted Green Synthesis of 3D Hierarchical Flower-Shaped $NiCo_2O_4$ Microsphere for High-Performance Supercapacitor. *ACS Appl. Mater. Interfaces* **2014**, *6* (3), 1773–1780.

CHAPTER 3

Global Abatement of Air Pollution Through Green Technology Routes

SIJO FRANCIS[1*], REMYA VIJAYAN[2], EBEY P. KOSHY[3], and BEENA MATHEW[4]

[1]*Department of Chemistry, St. Joseph's College, Moolamattom, India*

[2]*School of Chemical Sciences, Mahatma Gandhi University, Kottayam, India*

[3]*Department of Chemistry, St. Joseph's College, Moolamattom, India*

[4]*School of Chemical Sciences, Mahatma Gandhi University, Kottayam, India*

Corresponding author. E-mail: srsijofrancis@gmail.com

ABSTRACT

Air pollution seems to be the major unsolved problem of the modern era. This chapter deals with different causes of air pollutions and proposes some possible remedies. The green technological alternatives for the global abatement of air pollution are also suggested. The health problems associated with air contamination were also discussed.

3.1 INTRODUCTION

Environment pollution is a fruit of industrialization which shaped the modern lifestyle with the so-called use and throw culture. Nowadays the technological progress without effecting pollution is a risky game. Pollutants are substances emitted by industrial structures and products which are harmful to human beings and animals in particular and the environment at large. The proper functioning of the whole ecosystem is hindered by air pollution because the distribution of air pollutants does not obey boundary conditions since pollutants can travel long distances without any obstruction. Thus, air pollution is a serious ecological issue of the modern world. Consequently, countries began to think to pass pollution prevention act. The United States passed the pollution prevention act in 1990 for preventing the generation of pollutants and their abatement. International bodies such as the World Health Organization and others monitor air pollutions on a global basis.

3.1.1   CLASSIFICATION OF AIR POLLUTANTS

Causes of air pollution may be natural or human-made. Based on the origin, air pollutants are classified into primary pollutants and secondary pollutants.

3.1.1.1   PRIMARY POLLUTANTS

Primary pollutants are pollutants that are either directly emitted from natural sources or from anthropogenic activities. While natural activities such as volcanic eruptions, natural gas emissions, dust storms, seed germination, and marsh gas production are natural sources of primary pollutants; anthropogenic activities such as motor vehicle exhaust pipes and the outlet of factories contribute largely to the air pollution are manmade sources of air pollutants. Oxides of carbon (CO), oxides of sulfur (SO_x), oxides of nitrogen (NO_x), volatile organic compounds, and particulate matter are the main categories of primary pollutants.

3.1.1.2   SECONDARY POLLUTANTS

They are generated by the reaction of primary pollutants with other compounds. Ozone, acid rain, hydrocarbons, etc. are examples of secondary pollutants. Ozone is formed in stratosphere when primary pollutants such as hydrocarbon, NO_x, etc. react with atmospheric oxygen in the presence of sunlight.

3.1.2 CONTRIBUTORS TO AIR POLLUTANTS

Having seen different types of pollutants, let us see now what the contributors to air pollutants are. Modern means of transportation and automobiles are the main sources of pollution. Coal-burning in automobiles, chimneys of factories, machinery in factories, power plants, cooking in kitchens, building constructions, waste incinerators, forest fires, etc. are also contributing to air pollution. Air may be contaminated by different CO emitted through either incomplete or complete burning of fossil fuels. Greenhouse gases were supplied to the atmosphere by automobile engines. Minerals extraction by using large machines increases the amount of pollutants in air. NO_x were formed in the atmosphere usually from combustion processes and photochemical reactions in the atmosphere. Combustion of sulfur-containing compounds, volcanic activities, smelting of sulfide ores, sulfuric acid manufacturing unit, etc., contribute SO_x to the atmosphere. Bioaerosols are clouds of pollutants with the biological origin and are also a thread in urban areas. Metal particulate matter contains toxic heavy metals like Ni, Fe, Cd, Cr, Pb, etc., which also contribute to air pollution.

3.2 ENVIRONMENT IMPACTS

3.2.1 ACID RAIN

Acid rain is one of the environmental impacts resulting from the emission of NO_x and SO_x when air is washed out with water. Main contents of acid rain include sulfuric acid and nitric acid. Tones of agents of acid rain are expelled to air per year. Acid rain destroys monuments, buildings, marble statues, etc. Since, acid rain adversely affects the growth of plants and production of crops, the economy of the countries is largely affected by unexpected acid pouring. Besides vegetation and agro-industry, acid rain also affects fishery. Sulfuric acid content along with certain heavy metals destroys fish egg and thus offsprings of different kinds of fishes are destroyed in number and genus. This increases the acidic content in the river and ocean and thus the aquatic life is adversely affected. Acid rain is a great ecological threat that collapses the harmony of nature.

3.2.2   EUTROPHICATION

Eutrophication is the overabundance of plant nutrients especially in water bodies. It supports algae to bloom. Fertilizers washed away with soil erosion are mainly responsible for this phenomenon. NPK fertilizers mainly the nitrogenous fertilizers are responsible for this weed growth. Air-borne pollutants mainly sulfur and nitrogen-based pollutants contribute to dissolved nutrients for aquatic weeds. The dissolved oxygen content in water depletes seriously threatens aquatic life. Eutrophication is a natural process leading to aging of lakes. Human activities can greatly accelerate eutrophication in aquatic ecosystems. Emissions of nitrogen oxides from vehicles and other similar sources contribute to the amount of nitrogen entering aquatic ecosystems.[1]

3.2.3   INCORPORATION OF TOXINS INTO THE FOOD CHAIN

Biomagnification/bioamplification is the increase in the concentration of a toxin in the food chain from the tissues of tolerant organisms at a successively increasing level. For example, shellfish accumulate mercury in their body tissues and become a source of mercury contamination. Minamata accident is due to bioaccumulation of mercury in fish tissues. In the same way, airborne organic pollutants may also enter in multiple manners into the food chain.

3.2.4   OZONE HOLE

Ozone is O_3 which occurs in the stratosphere of the upper atmosphere. It shields our earth from UV-rays emitted from the sun. Ultraviolet A is responsible for photoaging and ultraviolet B is responsible for sunburn. Increase in air pollutants affects human skin adversely. Ultraviolet radiation, polynuclear aromatic hydrocarbons, easily volatile organic compounds, particulate matter, ozone, etc. affect human skin adversely. Exposure to ultraviolet radiation leads to skin aging and skin cancers. UV radiation increases the risk of skin diseases, cataracts, metabolic disorders, and decreases reproduction rates. Skin infection is mainly caused by oxidative stress mechanisms by losing antioxidant defense property of skin. Prolonged exposure of organic air pollutants has profound harmful effects on the skin. Cigarette smoke is also responsible for skin cancers and allergic skin conditions.

3.2.4.1   MONTREAL PROTOCOL

Montreal Protocol is an international agreement made in 1987. This is a result of serious discussions made by 28 countries at the Vienna convention for the protection of ozone layer. This protocol controls the production of chemicals that depletes ozone concentration in the atmosphere. Chloro-fluorocarbons from refrigerators, coolants, aerosol sprays produce chlorine atoms that are the major ozone-depleting gases. This treaty entreats countries to reduce the production of ozone-depleting gases.

3.2.5   PLANT GROWTH AND CROP PRODUCTION

Air pollutants like SO_x, NO_x, particulates, ozone, etc. have toxic effects on vegetation. The retardation of plant growth is caused by air pollution. The ozone exposure increased plant susceptibility to disease, pests, and environmental stresses. The crop and forest damage can also be caused by UV radiation resulted by ozone depletion. UV-rays damage wheat, cotton, soybean, and corn. These crops show reduced yields under exposure to ozone and thus reduce the crop production. Soil fertility and water level of soil are affected seriously by air pollution. Plant leaves are scavenged by airborne pollutants.[2]

3.2.6   ROAD SAFETY

Recent studies show that the air quality substantially also affects road safety. Road accidents cause material damages, bodily harm, and even to the loss of life. Air pollution may distract safe driving because many pollutants responsible involved in air pollution cause various harm to the driver, including the affecting the visibility of driving during the drive. The road vehicles themselves are the major source of air pollutants responsible for road accidents. Besides, the volume of vehicles on roads and road accidents rate are interrelated.

The sky darkening and reduction in visibility are the after-effects of smog and fog in industrial areas. Smog is a type of air pollution. Smog is formed in a warmer climate where the presence of UV radiation is high. *Photochemical smog originates from vehicle exhausts mainly from nitr*ic oxide and hydrocarbons as well as from other industrial sources. This also affects road safety. Haze, which constitutes a part of air pollution, is also a threat to road safety. Haze is produced by NO_x and SO_x combined with smoke from industries/wildfires. Haze-causing particles can be carried by the wind to a large distance. Thick blanket of haze particles affects photosynthesis by diminishing the availability of sunlight essential for it and reduce road safety. Thus, we can say that air quality is of at most importance in road safety. Reducing air pollution levels will decrease the rate of road accidents.[3]

3.2.7   CLIMATE CHANGE

Air pollutants such as greenhouse gases adversely affect the climatic conditions of the globe. Major greenhouse gases are carbon dioxide (CO_2) and methane (CH_4). CO_2 possesses the first position among the greenhouse effect. To control drastic climate change, greenhouse-gas emissions must be reduced. Reduction of fossil-fuel use is of the highest priority. CO_2 shoots up the atmospheric temperature. Increase in temperature which results in rising sea levels, extreme weather, and the increase of infectious and heat-related diseases are the main after-effects of global warming. Air pollution increases the risk for environmental catastrophes. It is seen that most of the flooded regions are rich in aerosols. The particulates, dust, and soot can produce intense rain, severe storms, and snowfall. According to the study by United States government, the particulate matter and the tiny solid pollutants coming from power plants can cause wildfires.

3.3   HEALTH EFFECTS

Since pollutants in the air are detrimental to human health, the public health is in crisis in cities. The Clean Air Act of United States Environmental Protection Agency (EPA) is to protect public health by regulating the emissions of these harmful air pollutants. The main diseases caused by air pollution include respiratory and cardiovascular-related issues. CO,

NO, and soot are the main pollutants detrimental to health. CO attacks hemoglobin and as a result carboxyhemoglobin is formed and the oxygen-binging capacity of blood gets reduced. NO can form additional compounds with hemoglobin and thus enter into the bloodstream. Soot is a particulate matter and contains tiny chemical particles, smoke, dust particles, or other allergens. Contents of smog and soot are almost similar. The smallest airborne particles in soot cause health problems including bronchitis, asthma, heart attacks, irritation to eyes and throat, chronic obstructive pulmonary disease, and even death. Besides the above-mentioned pollutants, toxic components such as polynuclear hydrocarbons which are exhausts from traffic exhausts and wildfire also cause lung irritation, liver issues, and even deadly cancer. In addition to the above man-made pollutants, there are also natural air pollutants. Some allergens from trees, grass, and weeds which are known as pollen and mold also cause air pollution. They are carried by air and are hazardous to health. Although no human activity is involved in carrying these allergens, they are also considered air pollution. Some health issues cause by air pollution are mentioned below.

3.3.1 BRAIN DISORDERS

Neurodegenerative diseases may be associated with low-quality air. Organic air pollutants like particulate matter, NO_x, SO_x, etc. affect brain structure and neuronal degeneration may happen which might eventually lead to Alzheimer or Parkinson disease. Exposure to ozone also influence mental health and lead to depressive episodes. Air pollution may lead to mood disorders, depressive disorders, and even to suicide.[4] Toxic air pollutants lead to mental health problems like anxiety.

3.3.2 BIRTH DISORDERS

Exposure to carbon monoxide and ozone may results in both birth disorders and abnormal birth. Premature mortality is happened by air pollutants. Anthropogenic sources of premature mortality by air pollution are agriculture, fossil fuel-fired by power plants, industry, biomass burning, and traffic. These contribute five out of six of air pollution while natural means contributes only one out of six.[5]

3.4   POLLUTION CONTROLLING STRATEGIES

The air pollution which is a social evil can be reduced by prevention and control. Legal bodies must ensure by routine monitoring the level of pollution emitted by vehicles. Quality of the fuel must be ensured. Alternate sources of energy are highly recommended. Since higher energy consumption, electricity generation, and an increase in the number of vehicles lead to an increase in the multiple pollutant emissions, moderate use of these will also reduce air pollution. As all air pollution control techniques concentrate on stationary sources with easy treatment at a relatively low cost.[6] The following biocompatible solutions may be recommended as low-cost means for the abatement of air pollution.

3.4.1   MODIFICATION OF AUTOMOBILE ENGINES—THERMAL OXIDATION

Designing automobile engines which emit minimal air pollutant when fossil fuels are burned is a current necessity. Automobiles are the main source of CO pollution and are mainly attributed to combustion under low stoichiometric ratio of fuel and air. Generally, fuel burn is completed at low temperatures and higher concentrations of oxygen. Internal engines are to be modified to reduce the production and emission CO to the atmosphere.

3.4.2   SUBSTITUTE FUELS

Substitutes for gasoline that produce fewer amount of CO are recommended. Substitute fuels for gasoline are highly recommended so that the amount of air pollutants on combustion may be reduced considerably. Unleaded petrol controls lead toxicity to a greater extent. Pyretic sulfur in coal must be removed and low-sulfur fuels are recommended.

3.4.2.1   BIOFUELS

Biofuels are fuels derived from biomass. Biodiesel, biogas, bioethanol, biomethanol, etc. are examples of biofuels. It reduces CO_2 emission and gives good fuel efficiencies. They are nontoxic and biodegradable too.

Biodiesel production involves transesterification reactions. Biodiesel is methyl esters. Diesel blended with 5% biodiesel is allowed in certain countries. Different crop plants are used for biodiesel production. Oil is generally extracted from their oilseeds. They include *Jatropha curcus*, *Pongamia pinnata*, *Calotropis gigantia*, *Euphorbia tirucalli*, etc. Degraded land can be cultivated by plantation of these trees. One advantage of biodiesel can be used in conventional diesel engines without modification. The major advantage of using biodiesel engines, however, is the remarkable reduction in the emission of air pollutants like CO_2, CO, SO_x, NO_x, hydrocarbon, particulate matter, etc. It can be used as a domestic fuel as well and is user-friendly. The production and the storage of biodiesel need to be promoted because it can be stored safely than fossil diesel. Since it is harmless to aquatic system, even marine stores could be used. Thus, biofuel is environmentally renewable and efficient in the abatement of air pollution.

Nanomaterials like calcium oxide and magnesium oxide have been used as biocatalyst in oil transesterification reactions for biodiesel production.[7] Biodiesel can be synthesized by transesterification of sunflower oil with methanol using nanocatalysts.[8] Ultrasonic-assisted transesterification using heterogeneous nanocatalysts for the production of biodiesel is a sustainable and environment-friendly approach.[9] Heterogeneous catalysts have high activity and high water tolerance properties.[10] Waste eggshell can also be used as a catalyst for biodiesel synthesis.[11] Microwave-assisted pyrolysis and solvolysis reactions are used for lignin conversion leading to renewable fuel on large-scale.[12]

3.4.2.2   BIOLUBRICANTS

Vehicle engines require lubricants. Biolubricants are biodegradable and possesses antiwear protection, and are nontoxic. Due to polar nature, natural oils like corn, soybean, olive, sunflower, peanut, etc. can be used as lubricants. Synthetic biodegradable lubricants include poly alpha olefins, diesters, polyglycols, etc. While esterification of oleic acid with small carbon chain alcohols yields biofuels, the same with long carbon chain alcohols yields biolubricants. Biolubricants can successfully be produced by enzyme-catalyzed esterification using ionic liquids.[13] The biolubricant isoamyl oleate can be produced by the enzymatic reaction in microfluidic reactor using Novozym 435 enzyme. The bioconversion is done in solvent-free media.[14] Synthesis of biolubricants may be extracted from effluents of palm oil mill enzymatic hydrolysis as well as noncatalytic esterification.[15] Many renewable hydrocarbon biolubricants with added qualities can be prepared from hydrodeoxygenation of vegetable oils at 50 bar pressure and 450 °C. This opens a low-cost and high-quality technology for the preparation of renewable hydrocarbons that can do the functionalities of biolubricants.[16]

3.4.2.3   HYDROGEN PRODUCTION

For clean development, hydrogen production from green, renewable, and sustainable energy resources are of higher priority. The splitting of water molecules into O_2 and H_2 gases by a technology is the basic chemistry of hydrogen production. The technologies of hydrogen production from fossil fuels like natural gas, coal, etc. are what so far known and used. The environment friendly and green methodologies include renewable resources like combustion of biomass, biomass pyrolysis, biohydrogen production using anaerobic bacteria, etc. Although, Solar-based hydrogen fuel production is expensive. Photovoltaic cell-based hydrogen production is also important and may lead to fast development in industry and commerce. Hydrogen production from supercritical water gasification (SCWG) of moisturized biomass is an effective thermochemical process. Highly moisturized biomass is also utilized directly in SCWG without any high-cost drying process.[17]

3.5   TREATMENT OF EXHAUST GASES

Treatment of exhaust gases prior to release to the atmosphere will decrease the amount of pollutants in the air. Some technologies used in this field are the following:

3.5.1   CATALYTIC CONVERTERS

Catalytic converters—two stages—remove NO_x emission from the atmosphere. First converter consists of Pt catalyst for the conversion of the emitted NO_x to N_2/ NH_3. Second converter facilitates an oxidation atmosphere for the efficient conversion of hydrocarbons/CO to H_2O/ CO_2. NO is also responsible for the depletion of O_3 levels. Excess oxygen levels limit the levels of NO. Solid absorbents like activated charcoal, activated alumina, activated silica gel, and molecular zeolites can function as catalysts.

3.5.2   *SELECTIVE CATALYTIC REDUCTION*

NO_x and sulfur dioxide (SO_2) are the main air pollutants. Awareness to protect the environment is progressively increasing. Catalytic reduction is the preferred technology used for onshore flue gas purification. Catalysts are useful for cleaning exhaust gases and they must work efficiently under large volumes of oxygen and moisture.[18] Modern catalysts are based on nanotechnology and current candidate of catalytic reduction mostly nano-catalysts. Nanotechnology is a promising method to remove air pollutants. Nanotubes and nanoparticles can be used. Carbon nanotubes (CNTs) with cerium oxide (CeO_2) effectively remove NO_x.[19] Selective catalytic reduction of NO can be effected by CeO_2–CNTs combinations.[20] CeO_2 nanoparticles and CNTs can also be used to remove the harmful exhaust gas emissions from additives in Diesterol.[21] Titanic acid nanotubes can also have catalytic capacity to remove NO_x.[22] Cr–MnO_x mixed-oxide used to remove NO by oxidation of NO to nitrogen dioxide (NO_2).[23]

The control of particulate emissions from renewable bioresources like biomass boiler plants using different scrubber systems like a washing tower, a Venturi scrubber, bubble-column scrubber show that they have good removal efficiency.[24] Electro-scrubbing process is a green method-ology for the removal of pollutant mist particles like CH_3SH as a model air pollutant.[25]

3.5.3   *ADSORPTION*

Gaseous pollutants are adsorbed on solid adsorbents in a suitable container. Some adsorbents used and gaseous pollutants collected are given below:

TABLE 3.1   Solid Adsorbents for the Removal of Air Pollutants.

Pollutant	Adsorbent
NO_x	Zeolites, silica get
H_2S	Iron oxides
SO_2	Dolomite, alkaline alumina
Organic solvent vapors	Activated charcoal

3.5.4   ELECTROSTATIC PRECIPITATORS

Electrostatic precipitators working at very high potentials are able to remove pollutants in the precipitated form. This is based on the principle of acquiring charge by aerosol particles when an electric field is provided. It is given by the equation:

$$F = E \times q,$$

where, F = force, E = potential gradient, and q = electrostatic charge.

The charge acquired particles are attracted toward an electrostatic precipitator. It can removed large sized pollutants aluminum or stainless steel parallel plates. They use big filters or scrubbers to remove dust particles.

3.5.5   HIGH-GRAVITY TECHNOLOGY

The high-gravity technology is the effective removal of particulate matter and gas pollutants from industries. Particulate collection is based on size, shape, velocity, gravity, momentum, etc. It works as follows: the effluent gas is applied to a big chamber and by decreasing velocities the particulate matter gets settled. The basic characteristics of high-gravity technique are mass transfer and high micro-mixing efficiency. According to this technology, the multiple air pollutants like SO_x, NO_x, CO_2, particulate matter, etc. can be removed simultaneously with high efficiency. The benefits of the high-gravity technology are: it has zero secondary pollution, it is low cost, it requires only small space, etc. It can be scaled-up into large scale and the method is reliable. This high-gravity technology finds applications in integrated air pollution control in industries.[6]

3.5.6   SCRUBBERS

Generally it helps remove liquid, solid, or gas contaminants. Scrubbers or adsorbers containing suitable liquid, generally water, the adsorbent, helps to remove pollutants present in a stream of gas. Chemical scrubbers like limestone and citric acid absorb SO_x from the atmosphere. Wet collectors (scrubbers) are for the removal of fine particulates in a gas stream. Pollutant gases are effectively washed using atomized liquid. Marine air pollution has to be controlled using wet scrubbers containing electrolyzed seawater using a liquid, generally water, to eliminate particulate matter. Scrubbers contain several chambers and it is called scrubbing cloud of water. NO_x and SO_x from ship emissions can be removed by using wet scrubbers. Absorbents may be NaOH, Na_2SO_3, NH_4OH, $Mg(OH)_2$, and $Ca(OH)_2$. The most attractive feature of wet scrubbing is that it can be done in low ambient temperatures.[26]

3.5.7   NANOTECHNOLOGY

Greenhouse gases are mainly CO_2, CH_4, nitrous oxide, and fluorinated gases. The adsorption of air pollutants on nanomaterials is more efficient and cost-effective because of their high surface area, their adsorption capacity, and ability for regeneration. Solid adsorbents for CO_2 adsorption include calcium-based Nanomaterials, nanoparticles in combination with alkali metals, functionalized CNTs. For the decomposition of CH_4 and nitrous oxides (NO_x) metallic nickel nanoparticles, TiO_2 coated with stainless steel web net, titanate nanotubes and their derivatives etc. are suggested. Modified TiO_2–silver catalyst was suggested for the photodecomposition of (N_2O) into nitrogen and oxygen. For the removal of SO_2, activated carbon deposited on iron nanoparticles may be used as adsorbent. SO_2 adsorption process leads to some changes in the magnetism of magnetic nanoparticles.[27] Household air pollution can be minimized by LPG by replacing biomass, wood, straw, and dung etc. for household cooking purpose.

3.5.8 NONTHERMAL PLASMA PROCESSING (NTP)

Non thermal plasma is a kind of non-equilibrium plasma while in actual plasma state the components are in thermal equilibrium. Highly energetic electrons produce plasma state and the developed free radicals decompose the pollutants. NTP has ability to induce various chemical reactions at normal pressure and temperature. Generation of ozone for water disinfection and dust removal are the main benefits of NTP related to air pollution control. Hybrid NTP system has a combined effect of adsorbents and catalysts and can be used in industries.[28]

3.6   GREEN TECHNOLOGIES FOR AIR POLLUTION CONTROL

3.6.1   MEMBRANE SEPARATION

The membranes separation of gases pollutants requires more money than traditional separation methods.[29] Semi permeable but hollow-fiber membranes are used to separate polluting gases from atmosphere. The pore size of the membrane has a crucial role in the activity of membranes and because of smaller size, oxygen molecules passes through the *membrane* easily than *nitrogen* molecule. Environmental sustainability of the CO_2 capture based on membrane separation process depends on material of the membrane, thickness, area, the net power consumption etc.[30] For the removal H_2S and CO_2 gases, combined technology contributed the advantages of membrane and amine processes find more useful. Nano porous graphene membrane is beneficial for gas separation with high permeability and selectivity because of highly defined size and geometry of the pores.[31] The limitations of commercial membranes are neutralized to certain extent by polymeric (organic) membranes and they promote acid gas removal process intensified. Major acid gases are CO_2 and hydrogen sulfide.[32]

3.6.2   BIOSORPTION

The removal of substances from solution by bio-derived materials is known as biosorption. The materials used for the purpose are generally renewable materials and are known as biosorbents. Biosorbents include microorganisms like fungi, bacteria, and algae and polysaccharides like chitosan or proteins. This is an environmental remediation technique to remove toxins and harmful substances.[33] To remove organic contaminants from aquatic environments biosorption is generally used because of the availability of biosorption materials and efficiency.[34] Phenolic pollutants heavy metals, agricultural pollutants, etc. can be effectively removed.

3.6.3 BIOFILTRATION

Biofiltration of air pollutants is similar to biofiltration of industrial effluents. The pollutant gas is allowed to pass over a moist porous bed of some material that contains bacteria, fungus, or any other microorganisms. They must be maintained at ambient temperature, concentration of oxygen, and other nutrients.[35]

3.6.4 BIOCHAR SEQUESTRATION

Carbon sequestration refers to the process of storing carbon in soil and thus removing CO_2 from the atmosphere. It is an existing approach of removing carbon from the atmosphere. It is done by planting more trees to sequester CO_2 in biomass. Substantial withdrawal of CO_2, biochar sequestration technology is suggested.[36] Thus, along with this bioenergy production, reduction in emission is also attained. It has a crucial value in global carbon market. Biochar is the process of heating plant biomass in the absence of oxygen at high temperatures. It is actually a carbon-negative industry. Biochar improves the fertility and structure of soils by improving the biomass production. Biochar enhances efficiency and decreases their leach out.[37]

3.6.5 GREEN ROOFS

Plants inhale gaseous pollutants like particulates through their leaves and break down some organic compounds through their tissues. Plants decrease temperature at the surface and thus decrease the rate of photochemical reactions that entail the formation of ozone. In urban areas, space for the cultivation of forests is less and thus roadsides, parking space, etc. are being used for the same. To mitigate pollution and decrease the quantity of pollutants, the expensive but useful technology called green roofs are suggested in urban areas.[38] Green roof shades improve quality of air, decreases temperature and facilitate cooling effect, improves human health, and thus benefits the society not only the owner of the building. To maximize the quality of air, deciduous plant species are recommended. Levels of pollutants like NO_x, SO_x, particulate matter, and ozone can be fruitfully decreased by green rooftop technology. A green roof can keep carbon equilibrium of the atmosphere and has a specific role in maintaining quality of water. Green roofing has good life span (45 years) over conventional roofing (20 years). This decreases the frequency of pollution by building constructions. Green roof has an architectural and aesthetical value. The environment sustainable practice that decreases air pollution needs to be supported by the authorities. Management of green roofs, chance of contamination, and economical criteria are to be considered.[39]

3.6.6   HERBAL REMEDIES

Some herbal plants around us can help us to fight against the negative effects of air pollution mainly on human bodies. They have antihistamine and antioxidant qualities and are harmful to toxins. Lobelia contains alkaloid Lobeline, helps easy breathing. Leaves of the eucalyptus tree fight against cough and congestion. Lungwort was used for the treatment of respiratory illnesses. Oregano contains carvacrol and rosmarinic and has medicinal properties.[40]

3.6.7   GREEN INFRASTRUCTURE

The pollutants harmful in cities are NO_2, SO_2, ozone, and particulate matter and they cause pulmonary and cardiac diseases. Street-level concentrations of NO_2 and particulate matter in cities can be reduced by dispersion of them at the street-level by in-canyon air recirculation and dry deposition to surfaces. The Green infrastructure of these canyons increases the ability of vegetation and remove pollutants and improve air quality in urban cities.[41]

3.7 AIR QUALITY STANDARDS

Air quality monitoring is an important task related to the public human health and this shows the risk level of air pollution.[42] EPA concerned about six air pollutants namely NO_x, O_3, SO_2, CO, particulate matter, and lead. Air quality standards means the allowed level of pollutants in a specified period in a given geographic area in reference units of time and concentration. In USA, "sanitary standards" and USSR "hygienic standards" are used. Air quality monitoring stations situated at different parts played their eminent role in air quality management mission.

TABLE 3.2 Air Quality Standards in the United States.

Air pollutant	Air quality standards	Period	Measurement
So_x	80 µg/m³	Annual	Arithmetic mean
NO_2	100 µg/m³	Annual	Arithmetic mean
CO	10 µg/m³	24 h	Concentration
Particulate matter	75 µg/m³	Annual	Geographic mean

Air pollution can be assessed by collecting data using an efficient sampling technique. For example, "high volume sampler" is used to determine suspended particulate matter and "dust fall gar" is used to determine the settleable particles in the air.

3.8 CONCLUSION

Air pollution has no boundaries, and therefore control of air pollution at the source is of utmost importance. Countries establish new regulations to meet the needs of urbanization and modernization. The air pollution control mission has a prominent role in the economic growth of nations as well.

KEYWORDS

- **household**
- **petroleum**
- **industrial**
- **renewable**
- **economic**
- **global elimination**

REFERENCES

1. Kampa, M.; Castanas, E. Human Health Effects of Air Pollution. *Environ. Pollut.* **2008**, *151* (2), 362–367.
2. Franzaring, J.; van der Eerden, L. J. M. Accumulation of Airborne Persistent Organic Pollutants (POPs) in Plants. *Basic Appl. Ecol.* **2000**, *1* (1), 25–30.
3. Sager, L. Estimating the Effect of Air Pollution on Road Safety Using Atmospheric Temperature Inversions. *J. Environ. Econ. Manage.* **2019**, *98*, 102250.
4. Gładka, A.; Rymaszewska, J.; Zatoński, T. Impact of Air Pollution on Depression and Suicide. *Int. J. Occup. Med. Environ. Health* **2018**, *31* (6), 711–721.
5. Lelieveld, J.; et al. The Contribution of Outdoor Air Pollution Sources to Premature Mortality on a Global Scale. *Nature* **2015**, *525* (7569), 367.
6. Pan, S.-Y.; et al. Development of High-Gravity Technology for Removing Particulate and Gaseous Pollutant Emissions: Principles and Applications. *J. Cleaner Prod.* **2017**, *149*, 540–556.
7. Sekhon, B. S. Nanotechnology in Agri-Food Production: An Overview. *Nanotechnol. Sci. Appl.* **2014**, *7*, 31.
8. Varghese, R.; et al. *Ultrasonication Assisted Production of Biodiesel from Sunflower Oil by using CuO: Mg Heterogeneous Nanocatalyst.* IOP Conference Series: Materials Science and Engineering. Vol. 225. No. 1. IOP Publishing, 2017.
9. Xie, W.; Wang, J. Enzymatic Production of Biodiesel from Soybean Oil by Using Immobilized Lipase on Fe_3O_4/Poly (Styrene-Methacrylic Acid) Magnetic Microsphere as a Biocatalyst. *Energy Fuels* **2014**, *28* (4), 2624–2631.
10. Chouhan, A. P. S.; Sarma, A. K. Modern Heterogeneous Catalysts for Biodiesel Production: A Comprehensive Review. *Renewable Sustainable Energy Rev.* **2011**, *15* (9), 4378–4399.
11. Nasrollahzadeh, M.; Sajadi, S. M.; Hatamifard, A. Waste Chicken Eggshell as a Natural Valuable Resource and Environmentally Benign Support for Biosynthesis of Catalytically Active Cu/Eggshell, Fe_3O_4/Eggshell and Cu/Fe_3O_4/Eggshell Nanocomposites. *Appl. Catal. B* **2016**, *191*, 209–227.
12. Yunpu, W.; et al. Review of Microwave-Assisted Lignin Conversion for Renewable Fuels and Chemicals. *J. Anal. Appl. Pyrolysis* **2016**, *119*, 104–113.

13. Bányai, T.; et al. Biolubricant Production in Ionic Liquids by Enzymatic Esterification. *Hung. J. Ind. Chem.* **2011,** *39* (3), 395–399.
14. Madarász, J.; et al. Solvent-Free Enzymatic Process for Biolubricant Production in Continuous Microfluidic Reactor. *J. Cleaner Prod.* **2015,** *93*, 140–144.
15. Syaima, M. T. S.; et al. The Synthesis of Bio-Lubricant Based Oil by Hydrolysis and Non-Catalytic of Palm Oil Mill Effluent (POME) Using Lipase. *Renewable Sustainable Energy Rev.* **2015,** *44*, 669–675.
16. Ho, C. K.; McAuley, K. B.; Peppley, B. A. Biolubricants Through Renewable Hydrocarbons: A Perspective for New Opportunities. *Renewable Sustainable Energy Rev.* **2019,** *113*, 109261.
17. Hosseini, S. E.; Wahid, M. A. Hydrogen Production from Renewable and Sustainable Energy Resources: Promising Green Energy Carrier for Clean Development. *Renewable Sustainable Energy Rev.* **2016,** *57*, 850–866.
18. Sun, Y.; Zwolińska, E.; Chmielewski, A. G. Abatement Technologies for High Concentrations of NO_x and SO_2 Removal from Exhaust Gases: A Review. *Crit. Rev. Environ. Sci. Technol.* **2016,** *46* (2), 119–142.
19. Mei, D.; et al. Role of Cerium Oxide Nanoparticles as Diesel Additives in Combustion Efficiency Improvements and Emission Reduction. *J. Energy Eng.* **2015,** *142* (4), 04015050.
20. Chen, X.; Gao, S.; Wang, H.; Liu, Y.; Wu, Z. Selective Catalytic Reduction of NO Over Carbon Nanotubes Supported CeO_2. *Catal. Commun.* **2011,** *14* (1), 1–5.
21. Selvan, V. A. M.; Anand, R. B.; Udayakumar, M. Effect of Cerium Oxide Nanoparticles and Carbon Nanotubes as Fuel-Borne Additives in Diesterol Blends on the Performance, Combustion and Emission Characteristics of a Variable Compression Ratio Engine. *Fuel* **2014,** *130*, 160–167.
22. Sun, Y.; Zwolińska, E.; Chmielewski, A. G. Abatement Technologies for High Concentrations of NO_x and SO_2 Removal from Exhaust Gases: A Review. *Crit. Rev. Environ. Sci. Technol.* **2016,** *46* (2), 119–142.
23. Chen, Z.; et al. Cr–MnOx Mixed-Oxide Catalysts for Selective Catalytic Reduction of NO_x With NH_3 at Low Temperature. *J. Catal.* **2010,** *276* (1), 56–65.
24. Bianchini, A.; et al. Performance Analysis of Different Scrubber Systems for Removal of Particulate Emissions from a Small Size Biomass Boiler. *Biomass Bioenergy* **2016,** *92*, 31–39.
25. Govindan, M.; Moon, I.-S. Uncovering Results in Electro-Scrubbing Process Toward Green Methodology During Environmental Air Pollutants Removal. *Process Saf. Environ. Prot.* **2015,** *93*, 227–232.
26. Yang, S.; et al. Removal of NO_x and SO_2 from Simulated Ship Emissions Using Wet Scrubbing Based on seawater Electrolysis Technology. *Chem. Eng. J.* **2018,** *331*, 8–15.
27. Ibrahim, R. K.; et al. Environmental Application of Nanotechnology: Air, Soil, and Water. *Environ. Sci. Pollut. Res.* **2016,** *23* (14), 13754–13788.
28. Kim, H.-Ha. Nonthermal Plasma Processing for Air-Pollution Control: A Historical Review, Current Issues, and Future Prospects. *Plasma Process. Polym.* **2004,** *1* (2), 91–110.

29. Zhang, X.; et al. Grafted Multifunctional Titanium Dioxide Nanotube Membrane: Separation and Photodegradation of Aquatic Pollutant. *Appl. Catal. B* **2008,** *84* (1–2), 262–267.

30. Giordano, L.; Roizard, D.; Favre, E. Life Cycle Assessment of Post-Combustion CO_2 Capture: A Comparison Between Membrane Separation and Chemical Absorption Processes. *Int. J. Greenhouse Gas Control* **2018,** *68,* 146–163.

31. Sun, C.; Wen, B.; Bai, B. Application of Nanoporous Graphene Membranes in Natural Gas Processing: Molecular Simulations of CH_4/CO_2, CH_4/H_2S and CH_4/N_2 Separation. *Chem. Eng. Sci.* **2015,** *138,* 616–621.

32. Razavi, S. M. R.; Shirazian, S.; Najafabadi, M. S. Investigations on the Ability of Di-Isopropanol Amine Solution for Removal of CO_2 from Natural Gas in Porous Polymeric Membranes. *Polym. Eng. Sci.* **2015,** *55* (3), 598–603.

33. Fomina, M.; Gadd, G. M. Biosorption: Current Perspectives on Concept, Definition and Application. *Bioresour. Technol.* **2014,** *160,* 3–14.

34. Chaukura, N.; et al. Biosorbents for the Removal of Synthetic Organics and Emerging Pollutants: Opportunities and Challenges for Developing Countries. *Environ. Dev.* **2016,** *19,* 84–89.

35. Abdo, P.; Huynh, B. P.; Irga, P. J.; Torpy, F. R. Evaluation of Air Flow Through an Active Green Wall Biofilter. *Urban Forestry Urban Greening* **2019,** *41,* 75–84.

36. Field, C. B.; Mach, K. J. Rightsizing Carbon Dioxide Removal. *Science* **2017,** *356* (6339), 706–707.

37. Yargicoglu, E. N.; et al. Physical and Chemical Characterization of Waste Wood Derived Biochars. *Waste Management* **2015,** *36,* 256–268.

38. Ragheb, A.; El-Shimy, H.; Ragheb, G. Green Architecture: A Concept of Sustainability. *Procedia Social Behav. Sci.* **2016,** *216,* 778–787.

39. Shafique, M.; Kim, R.; Rafiq, M. Green roof benefits, opportunities and challenges–A review. *Renewable Sustainable Energy Rev.* **2018,** *90,* 757–773.

40. Oualili, H.; Nmila, R.; Chibi, F.; Lasky, M.; Mricha, A.; Rchid, H. Chemical Composition and Antioxidant Activity of Origanum Elongatum Essential Oil. *Pharmacognosy Res.* **2019,** 11(3), 283..

41. Pugh, T. A. M.; et al. Effectiveness of Green Infrastructure for Improvement of Air Quality in Urban Street Canyons. *Environ. Sci. Technol.* **2012,** *46* (14), 7692–7699.

42. Qiao, X.; et al. Evaluation of Air Quality in Chengdu, Sichuan Basin, China: are China's Air Quality Standards Sufficient Yet?. *Environ. Monitor. Assess.* **2015,** *187* (5), 250.

CHAPTER 4

Application of Green Technology in Water and Wastewater Treatments

REMYA VIJAYAN[1*], SIJO FRANCIS[2], and BEENA MATHEW[1]

[1]*School of Chemical Sciences, Mahatma Gandhi University, Kottayam, India*

[2]*Department of Chemistry, St. Joseph's College, Moolamattom, India*

Corresponding author. E-mail: remyavijayan88@gmail.com

ABSTRACT

Wastewater originated from various industries like textile, agriculture, food, petrochemical, polymer, pharmaceutical, etc. and contains a large number of contaminants of oil and salt of inorganic and organic compounds. When this wastewater released into the ecosystem without any appropriate treatments causes major ecological issues with high environmental impacts. Also, the natural fresh water resources are getting depleted because of the increased demand for fresh water supply. There are different physical, chemical, and biological methods are developed for the treatment of water but these methods cannot abolish the contaminants. And also most of these conventional methods are very expensive. The development of green technology for water treatments has received enormous interest over recent years due to its significant advantages to the environment, society, and economy. In this chapter, we discuss the various green technologies for the treatment of water and wastewater.

4.1 INTRODUCTION

The world's population has increased to more than seven billion people. Each year world population continues to increase with a 1.2% growth rate.

This results in the increased demands on water purity. Population increase, climate variations, and fast development of several nations and the subsequent large usage of water and contamination of water resources have raised anxiety regarding the unsustainability of existing water use patterns and supply methods. In the perception of the total hydrologic cycle, the adequate global amount of water is normally accessible for the present population but the concentration of world water resources in particular regions is resulting in the emergence of severe water deficiencies in other places.[1]

Inappropriately handled freshwater resource systems also cause significant water pollution problems other than water scarcity. Around 90–95% of untreated urban sewage in the developing countries released directly into surface waters without any purification process and water regulations. Consequently, water pollution problems are high in developing countries than in other countries.[2] The contamination of water resources dangerously affects the quality and supply of freshwater, mainly for domestic and industrial purposes. To overcome these situations, awareness of environmental responsibilities is needed.[3] It is very essential to develop sustainable wastewater treatment methods for confirming clean water and energy accessibility for upcoming generations. However, the selection and implementation of suitable cost-effective and environmental friendly treatment procedures are very essential. There are different conventional techniques are developed for the purification of water to a desirable quality.

Water treatment involves the removal of unwanted chemical, physical, and biological pollutants from raw or contaminated water to produce pure water for specific applications like human consumption, medical, industrial, chemical, and pharmacology requirements. It is not possible to recognize a water sample of fine quality by visual observation. The only method to attain the details required for deciding the suitable technique for water treatment is chemical analysis. The chemical analysis is somewhat expensive. The international standards like the World Health Organization (WHO) or governments are usually set the standards for drinking water quality.

The World Health Organization report (WHO 2007) in 2007 says that 1.1 billion people lack access to an improved drinking water supply. Yearly four billion diarrheal disease cases are reported of which 88% are originated from unsafe water and insufficient sanitation and cleanliness. Each year around 1.8 million people die due to diarrheal diseases. In many developing countries one of the main public health aims is the reduction of deaths caused by waterborne diseases. According to WHO 2005 reports, the modification of the environment by providing safe drinking water reduces 94% of diarrheal issues. Implementation of green technology at home for treating water like chlorination, filters, and solar disinfection, and keeping of water in good containers could save a large number of lives each year.[4]

There are different conventional methods like physical, chemical, mechanical, or biological or in some cases, the combination of these methods is used for the treatment of wastewater. The main aim of water treatment is to remove any solids and organic matter from the water. In physical treatment methods, the waste materials are separated or isolated from the mainstream and it does not involve any degradation of waste material. While in biological treatments, the microbes are used to feed on the organic waste and pH and aeration are adjusted to maintain microbial activities. The development of green technology for water treatments has received enormous interest over recent years due to its significant advantages to the environment, society, and economy. In this chapter, we discuss the various green technologies for the treatment of water and wastewater.

4.2 GREEN TECHNOLOGY METHODS FOR WATER TREATMENTS

The main purposes of water and wastewater treatment green technologies are:

(i) to decrease and preserve the exploitation of water and related nonrenewable energy resources
(ii) to avoid pollution and mishandling of water and other natural resources
(iii) to keep biodiversity, habitats, and ecologies, and
(iv) to make sure that upcoming generations can meet up their own requirements.

An environment friendly approach is required to overcome the consequences of the use of toxic chemicals and solvents in water treatment methods. Some of the green technology methods for water treatments are given below.

4.2.1   ADVANCED OXIDATION PROCESS (AOPS)

AOPs have been defined by Glaze et al.[5] and it involves several oxidation steps and is used to removing organic compounds in water by a set of the chemical treatment process. These methods are better substitutes for the removal of dissolved recalcitrant organic substances, which would not be completely eradicated by conventional methods. In this process, highly reactive intermediates like OH radicals are produced by the following routes.[6]

- Oxidation with O_2. Here the reaction is carried out in a temperature range between ambient conditions and those found in incinerators.

For example, wet air oxidation (WAO) processes (1–20 MPa and 200–300°C)

- Use of ozone and H_2O_2 and/or photons (high energy oxidants) for the generation of OH radicals

The OH radicals are very reactive and non-selective chemical oxidants. Once produced, it attacks almost all organic compounds. It can degrade the noxious substances present in the wastewater. Due to the high oxidation potential of OH radicals ($E° = 2.8$ V), it can react with all types of organic compounds results in complete mineralization of these compounds by the formation of water, carbon dioxide, inorganic salts or transfer it into less aggressive products.[7–9] AOPs reduce the concentration of pollutants from a few hundred ppm to less than 5ppb.[10] Destruction of pollutant substances and the prevention of subsequent formation of toxic residue are some of the important benefits of AOPs. But in conventional nondestructive physical separation processes like flotation, filtration, and adsorption with active coal only eliminate the pollutants and transferring it into other products.[11] The AOP uses strong oxidizing agents like hydrogen peroxide, Fenton's reagents, and ozone to generate hydroxyl radicals. Sometimes these reagents are used with ultraviolet (UV) light which enhances the reaction.

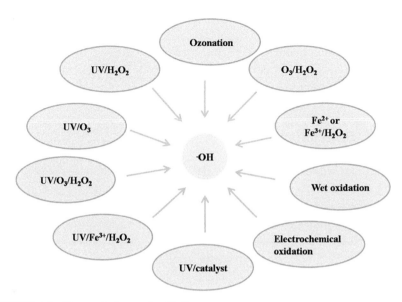

FIGURE 4.1   Types of advanced oxidation processes.

The AOP can be divided into two categories on the basis of whether UV light is used for the reaction.[12] These are given in Table 4.1.

TABLE 4.1   Classification of the Advanced Oxidation Process.

Photochemical	Non-photochemical
UV/H_2O_2	Ozonation
UV/O_3	Ozonation with hydrogen peroxide (O_3/H_2O_2)
UV/H_2O_2/O_3	Fenton (Fe^{2+} or Fe^{3+}/H_2O_2)
Photo-Fenton (Fe^{3+}/H_2O_2/UV)	Wet oxidation (WO)
Photocatalytic oxidation (UV/catalyst)	Electrochemical oxidation

Of these different AOPs, the most used methods for the removal of toxic organic contents in the wastewater are ozonation, photocatalytic degradation, Fenton's reagent (H_2O_2/Fe^{2+}), Photo-Fenton, and WAO. These methods are also very effective in the removal of organic dyes from the water media. The combinations of these different methods are also used for effective removal.

4.2.2 *NONPHOTOCHEMICAL ADVANCED OXIDATION PROCESSES*

In the nonphotochemical AOP, the hydroxyl radicals are generated via different methods without using light radiation. The important nonphoto-chemical AOPs are given below.

4.2.2.1 *OZONATION*

Ozone has a high reduction potential (2.07 V), and it can react with organic compounds directly. Nowadays, the ozonation is extensively used for the water treatment because this reaction does not produce any chlorinated products that may generate in the chlorine disinfection process.[13] Ozone is not harmful to any organisms and ozone-based AOPs are very ecofriendly. The ozonation methods are broadly applied for the removal of organic pollutants particularly the colored compounds. The conjugated double bond present in the chromophore groups of colored compounds is easily cleaved by ozone to smaller molecules, either directly, or indirectly.[14,15]

The ozone is decomposed to give hydroxyl radicals according to the following equation. This reaction is catalyzed by hydroxyl ions (OH$^-$)

$$2O_3 + 2H_2O \longrightarrow 2O\cdot H + 2HO_2\cdot + O_2$$

The ozonation processes take place in the following steps:

- Formation of ozone
- Dissolution of ozone in wastewater
- Oxidation of organic compounds

4.2.2.2 *OZONATION WITH HYDROGEN PEROXIDE (O_3/H_2O_2)*

Hydrogen peroxide (H_2O_2) is a readily obtainable oxidant and is less expensive. When H_2O_2 is added to the zone, the decomposition cycle of ozone initiated and OH radicals are formed.[16]

H_2O_2 reacts with ozone when present as an anion, HO_2^-.

$$H_2O_2 \longrightarrow HO_2^- + H^+$$

$$HO_2^- + O_3 \longrightarrow HO_2{}^{\cdot} + \mathbf{O_3^{\cdot-}}.$$

The reaction is continuous by the indirect pathway as illustrated above and OH radicals are generated.[17] From the analysis of different reaction steps, it is concluded that two ozone molecules produce two OH radicals:

$$2O_3 + H_2O_2 \longrightarrow 2O^{\cdot}H + 3O_2.$$

The study about the removal of the atrazine in filtered Seine River water showed that the degradation rate of pesticide is higher when water was treated with ozone-hydrogen peroxide mixture as a contrast to ozone only. The optimum H_2O_2/O_3 mass ratio was from 0.35 to 0.45. The rates of degradation are affected by factors like ozone dosage, contact time, and alkalinity of water.[18]

4.2.2.3   FENTON'S REAGENT (H_2O_2/FE^{2+}) OXIDATION

Fenton's reagent oxidation is a catalytic oxidation process, which involves a mixture of strong chemical oxidizer (hydrogen peroxide), ferrous ions as a catalyst, and an acid as an optimum pH adjuster. The Fenton process was reported by Fenton for maleic acid oxidation.[19] Fenton's process is an easy way to generate hydroxyl radicals without any special apparatus and chemicals and takes place at ambient temperature and pressure. This is a simple method for oxidation, as hydrogen peroxide and iron salts are readily obtainable, easy to handle, and environmentally benign.[20] The organic compounds are destructed by reacting with OH radicals.

In acidic medium, the reaction between the hydrogen peroxide (H_2O_2) and ferrous ions Fe(II) leads to the formation of a hydroxyl ion and a hydroxyl radical by decomposition of H_2O_2, and the oxidation of Fe(II) to Fe(III). This can be represented by the equation as follows:

$$Fe^{2+} + H_2O_2 \longrightarrow Fe^{3+} + O^{\cdot}H + OH^-$$

In the presence of excess H_2O_2, the Fe (II) oxidizes to Fe (III) within a few seconds and the rate constant for the reaction between ferrous ion and H_2O_2 is very high. Then, Fe (III) catalyzes the decomposition of H_2O_2 to hydroxyl radicals.

$$Fe^{3+} + H_2O_2 \rightleftharpoons H^+ + Fe\text{-}OOH^{2+}$$

$$Fe\text{-}OOH^{2+} \longrightarrow HO_2{\cdot} + Fe^{2+}$$

$$Fe^{2+} + H_2O_2 \longrightarrow Fe^{3+} + O{\cdot}H + OH^-$$

Hence, the Fenton's reagent catalyzed waste destruction is simply a Fe(III)-H_2O_2 system catalyzed process. Fenton's reagent with excess H_2O_2 is fundamentally a Fe (III)-H_2O_2 process (known as a Fenton-like reagent). Therefore, the ferrous ion in Fenton's reagent is able to change with the ferric ion. The iron salt is used as a catalyst and it is regenerated. Fenton's reagent can destroy phenols, nitrobenzene, and herbicides present in water, in addition, to decrease chemical oxygen demand (COD) in wastewater.[21–24]

4.2.2.4   WET OXIDATION (WO)

WO is a destructive and environmentally secure technology for treating water containing organic pollutants.[25] The WAO process was first patented by Zimmerman. By this technique, pollutants present in the water are removed by oxidation with an oxidant such as oxygen or air under high-pressure conditions (10–220 bar) and high temperatures (150–370°C), leading to the formation of hydroxyl radicals.[26] This technology is generally named WO when pure oxygen is used and WAO when air is supplied to the system. When hydrogen peroxide is used as an oxidant instead of oxygen it is known as wet peroxidation (WPO).

By this process, the organic pollutants are not fully removed but are converted into intermediate end products with a considerable decrease in COD and the total organic carbon. The last aqueous effluent will contain a considerable quantity of low molecular weight organics, ammonia, inorganic acids, and inorganic salts which are highly biodegradable than the untreated effluents. The efficiency of WAO can be enhanced with the presence of carbon materials, noble metals (Ru, Rh, Pd, Ir, Pt, etc.), and oxides of Cr, Mn, Fe, Co, Ni, Cu, Zn, and Mo.[27,28] This type of reaction is known as catalytic WAO. Agro-food streams, pulp and paper mill effluents, and leachates from solid waste have been treated by WO. During the WO process, the organic compounds are reduced to CO_2 or other harmless components; nitrogen is converted into NH_3, NO_3, or elementary nitrogen. The halogen compounds and sulfurs are changed into halides and sulfates. Also, dioxides or other harmful products like NOx, SO_2, HCl are not formed during this process.[29]

4.2.2.5   ELECTROCHEMICAL OXIDATION

There are different types of electrochemical advanced oxidation methods developed to degrade or mineralize the organic pollutants present in water. These are very environmental friendly methods and can produce electro-generated *in situ* hydroxyl radicals (OH). These radicals are very reactive and are powerful oxidizing agents ($E^{\circ}_{(OH/H2O)}$ = 2.8 V/SHE at 25°C).

Anodic oxidation is the simplest and well-known electrochemical oxidation method for water purification. In this method, organic pollutants present in the contaminated water are oxidized by heterogeneous hydroxyl radicals M(·OH) formed by direct charge transfer at the anode (M) as intermediate of O_2 evolution reaction from the oxidation of water as follows:

$$M + H_2O \longrightarrow M(O\cdot H) + H^+ + e^-$$

The use of active lower O_2 evolution overpotential anode (e. g., Pt/IrO_2) allows the conversion of organics into carboxylic acids since M(·OH) is chemisorbed on the anode surface with less oxidizing power. While the use of non-active high O_2 evolution overpotential anode (e. g., PbO_2/boron-doped diamond (BDD)) resulted in the formation of more reactive physisorbed M(.OH) with high oxidation power. These radicals can oxidize hardly oxidizable compounds like short-chain carboxylic acids.[30–32]

4.2.3   PHOTOCHEMICAL ADVANCED OXIDATION PROCESSES

The oxidation of organic pollutants by conventional methods like ozone or hydrogen peroxide oxidation does not completely oxidize them to CO_2 and H_2O in several circumstances.[33] Sometimes more toxic intermediate oxidation products remain in the solution. The termination of these oxidative reductions is attained by supplementing the reaction with UV radiation. Some of these methods are given below.

4.2.3.1   PHOTOCATALYTIC OZONATION (UV/O$_3$)

Ozonation combined with UV radiation is more effective to remove organic pollutants than ozonation only. The UV radiation enhances the decomposition of ozone and hence large numbers of OH radicals are formed thereby increases the rate of ozonation reaction.[34]

Ozone absorbs UV radiation at 254 nm wavelength. Firstly, an intermediate H_2O_2 is formed and finally hydroxyl radical is formed by the dissociation of H_2O_2.

$$O_3 + h\nu \longrightarrow O_2 + O\,(^1D)$$

$$O\,(^1D) + H_2O \longrightarrow H_2O_2 \longrightarrow 2O\cdot H$$

Hence, the ozone decomposition reaction and photocatalytic ozone decomposition reactions only differ in their initiation step. In photocatalytic ozone decomposition, the starting radical is formed photochemically by an electron transfer from photocatalysts to oxygen and not by the reaction of OH^- ion with ozone.[35] The complete mineralization of organic compounds with short molecular chains can be achieved by the UV/O$_3$ system.[36] Peyton et al. showed the higher effectiveness of the O$_3$/UV system for the exclusion of tetrachloroethane (C_2Cl_4) from water contrast to ozonation and photolysis only.[37]

4.2.3.2   ULTRAVIOLET IRRADIATION AND HYDROGEN PEROXIDE (UV/H$_2$O$_2$)

In the UV/H$_2$O$_2$ system, hydrogen peroxide (H_2O_2) is added in the presence of UV light to generate hydroxyl (OH) radicals. The H_2O_2 is a strong reducing agent it is used to remove low-level pollutants in wastewater.[38] Sometimes H_2O_2 alone is not an efficient oxidant to oxidize complex pollutants. When it is used with other reagents and energy sources which are able to dissociate it into free radical, its activity is increased. Normally, low or medium pressure mercury lamps are used for the photolysis of H_2O_2. When H_2O_2 irradiated with UV radiation of wavelength less than 300 nm, it gives OH radicals as follows.

$$H_2O_2 \xrightarrow{hv} 2OH^·.$$

The H_2O_2 can react with hydroxyl radical and the intermediary product formed. The reactions can be written as simple form as follows[39]:

$$H_2O_2 + OH^· \longrightarrow HO_2^· + H_2O$$

$$H_2O_2 + HO_2^· \longrightarrow OH^· + H_2O + O_2$$

$$2OH^· \longrightarrow H_2O_2$$

$$2HO_2^· \longrightarrow H_2O_2 + O_2$$

$$OH^· + HO_2^· \longrightarrow H_2O_2 + O_2.$$

The organic compounds are attacked by both hydroxyl radical and hydroperoxy radicals. But the reduction potential of hydroperoxy radicals (1.7 V) is lower than OH radical (2.8 V); hence, the formation of hydroperoxy radicals radical is not interested in these processes.

One of the main advantages of using the UV/H_2O_2 system is that the H_2O_2 is readily soluble in water, there is no mass transfer limitation, and it is a good source of hydroxyl radicals. Also, there are no separation techniques needed after treatment[40,41] UV/H_2O_2 system is very useful for the removal of organic dyes, particularly the azo dyes,[42] chlorophenols,[43] and other chlorinated compounds.[44] The complete mineralization atrazine, desethylatrazine, and simazine to carbon dioxide within reasonable irradiation times is achieved by using the UV/H_2O_2 system.[45]

4.2.3.3   OZONE–HYDROGEN PEROXIDE–UV RADIATION ($O_3/H_2O_2/UV$)

The addition of H_2O_2 to the UV/O_3 system enhances the decomposition of ozone, thus the rate of formation of OH radicals is increased.[46]

Trapido et al. proved that $O_3/H_2O_2/UV$ was the most effective system for the degradation of nitrophenol than O_3, O_3/H_2O_2, O_3/UV systems.[47] This system is also applied for the removal of volatile organic contaminants like benzene, acetone, dichloroethane, tetrachloroethane, etc. from groundwater.[48]

4.2.3.4 PHOTO-FENTON SYSTEM

In Photo-Fenton oxidation, Fe^{3+} is added to the H_2O_2/UV process. At acidic condition (pH= 3) $Fe(OH)^{2+}$ is formed.

$$Fe^{3+} + H_2O \longrightarrow Fe\,(OH)^{2+} + H^+$$

$$Fe\,(OH)^{2+} \rightleftharpoons Fe^{3+} + OH^-.$$

This complex is decomposed into Fe^{2+}, and OH ions when exposed to UV radiation.

$$Fe\,(OH)^{2+} \xrightarrow{\;hv\;} Fe^{2+} + H\cdot.$$

The organic pollutants can be completely removed by the Photo-Fenton process. Pignatello et al. demonstrated the complete mineralization of a number of herbicides and pesticides using the Photo-Fenton process. In another study, this process is used for the mineralization of chlorophenol.[49]

The increased efficiency of Photo-Fenton reactions can be attributed to the following reasons:

Photo-reduction of ferric ion: The ferrous ions are produced by the irradiation of ferric ion or ferric hydroxide. Then the cycle continues by the reaction of this ferrous ion with H_2O_2 and producing second hydroxyl radical and ferric ion.

Efficient use of light quanta: The absorption spectrum of ferric ion or hydroxyl ferric ions expands to the near-UV/visible region with a moderately large extinction coefficient. Hence, it is possible to carry out photo-oxidation and mineralization even by visible light.

4.2.3.5 PHOTOCATALYTIC OXIDATION

Photocatalytic oxidation is one of the most efficient and promising AOPs in which the organic compounds are fragmented into the water, carbon dioxide, and mineral salts. The activated species (hydroxyl radicals and superoxide radicals) are used for the complete mineralization.[50] The photocatalytic degradation by using semiconductors (TiO_2, ZnO, Fe_2O_3, WO_3, and CdS) is widely used for the degradation of organic pollutants. The nano-titanium dioxide (TiO_2, anatase form) is a commonly used photocatalyst due to its availability, chemical stability, extraordinary photocatalytic activity, nontoxicity, low cost, resistance to photocorrossion, optical and electrical properties, etc.[51,52] It is possible to use sunlight in the case of TiO_2 since it absorbs wavelength below 400 nm.

4.3 MEMBRANE BIOREACTOR (MBR) TECHNOLOGY

In recent years, MBR technology becomes one of the promising approaches for the treatment of wastewater and it gives enormous progress in wastewater treatments. MBR is the acronym for water and wastewater treatment processes which integrates a biological process with a membrane separation step. In general, membrane filtration is aimed at retaining biomass and other suspended materials so as to produce a clarified and disinfected permeate. First applications at full scale of MBRs were developed for treating shipboard wastewater in the late 1960s.[53,54]

The bioreactor acts as a biological treatment processor and the membrane is used as a filter in the filtration process. MBR technology has several environmental advantages like a smaller footprint, pathogen removal capacity, reduced sludge production, avoiding the use of chemicals for disinfection, superior separation efficiency, and greatly improved effluent quality compared to conventional activated sludge (CAS) process.[55,56] In MBR, with the help of ultrafiltration membranes, some kind of viruses can be retained. In contrast to the CAS process, the MBR technology can retain small molecular weight organic micropollutants.[57]

4.4 CONCLUSIONS

Nowadays, the execution of clean, eco-friendly, less energy, and waste generating methods and technologies is recognized with growing attention. So as to offer sustainable development, environment friendly substances and innovative green chemistry technologies should be exploited. Innovation and application of novel cleaner technologies will help to protect natural resources for the society and for upcoming generations. In this chapter, some green technology methods for wastewater treatment are discussed.

KEYWORDS

- **green technology**
- **wastewater**
- **cost-effective**
- **coagulation**
- **advanced oxidation**
- **membrane bio reactor**

REFERENCES

1. Pimentel, D.; Pimentel, M. *Human Population Growth. Encyclopedia of Ecology*; Elsevier: Amsterdam, Netherlands, 2008; pp 1907–1912.
2. Pimentel, D.; Whitecraft, M.; Scott, Z. R.; Zhao, L.; Satkiewicz, P.; Scott, T. J.; Phillips, J.; Szimak, D.; Singh, G.; Gonzalez, D. O.; Kyne, D. *Ethics of a Sustainable World Population in 100 Years. Encyclopedia of Applied Ethics*, 2nd ed.; Elsevier: London, 2012; pp 173–177.
3. Tseng, M. L.; Raymond, Tan R.; Anna Bella, S. M. Sustainable Consumption and Production for Asia: Sustainability through Green Design and Practice. *J. Clean. Prod.* **2013,** *40*, 1–5.
4. Show, K. Y. Green Technology. Department of Environmental Engineering. 2010.
5. Glaze, W. H.; Kang, J. W.; Douglas, H. C. The Chemistry of Water Treatment Processes Involving Ozone, Hydrogen Peroxide Ultraviolet Radiation. *Ozone Sci. Eng.* **1987,** 335–352.
6. Munter, R. Advanced Oxidation Processes-Current Status and Prospects. *Proc. Estonian Acad. Sci. Chem.* **2001,** *50*(2), 59–80.

7. Pera-Titus, M.; García-Molina, V.; Angel Baños, M.; Giménez, J. Degradation of Chlorophenols by Means of Advanced Oxidation Processes: A General Review. *Appl. Catal. B. Environ.* **2004,** *47*(4), 219–256.

8. Andreozzi, R.; Caprio, V.; Insola, A.; Marotta, R. Advanced Oxidation Processes (AOP) for Water Purification and Recovery. *Catal. Today* **1999,** *53*(1), 51–59.

9. Bolton, J. R.; Bircher, K. G.; Tumas, W.; Tolman, C. A. Figures-of-Merit for the Technical Development and Application of Advanced Oxidation Processes. *J. Adv. Oxid. Technol.* **1996,** *1*(1), 13–17.

10. Masroor, M.; Mehrvar, M.; Ein-Mozaffari, Farhad. An Overview of the Integration of Advanced Oxidation Technologies and Other Processes for Water and Wastewater Treatment. *Int. J. Eng.* **2009,** *3*(2), 120–146.

11. Crittenden, J. C.; Suri, R. P. S.; Perram, D. L.; Hand, D. W. Decontamination of Water using Adsorption and Photocatalysis. *Water Res.* **1997,** 31(3), 411–418.

12. Mota, A. L. N.; Mota, A. L. N.; Albuquerque, L. F.; Beltrame, L. T. C.; Chiavone-Filho, O.; Machulek Jr,, A.; C. A. O. Nascimento. Advanced Oxidation Processes and their Application in the Petroleum Industry: A Review. *Braz. J. Pet. Gas* **2009,** *2*(3), 122–142.

13. Augugliaro, V.; Litter, M.; Palmisano, L.; Soriad, J. The Combination of Heterogeneous Photocatalysis with Chemical and Physical Operations: A Tool for Improving the Photoprocess Performance. *J. Photochem. Photobiol. C. Photochem. Rev.* **2006,** *7*(4), 127–144.

14. O'Shea, K. E.; Dionysios, D. D. Advanced Oxidation Processes for Water Treatment. *J. Phys. Chem. Lett.* **2012,** *3,* 2112–2113.

15. Palit, S. An Overview of Ozonation Associated with Nanofiltration as an Effective Procedure in Treating Dye Effluents from Textile Industries with the Help of a Bubble Column Reactor. *Int. J. Chem. Sci.* **2012,** *10*(1), 27–35.

16. Gottschalk, C.; Libra, J. A; Saupe, A. *Ozonation of Water and Waste Water*; WILEY-VCH, 2000.

17. Hoigne, J. Mechanisms, Rates and Selectivities of Oxidations of Organic Compounds Initiated by Ozonation of Water. In *Handbook of Ozone Technology and Applications;* Ann Arbor Science Publ.: Ann Arbor, MI, 1982.

18. Paillard, H. Optimal Conditions for Applying an Ozone-Hydrogen Peroxide Oxidizing System. *Water Res.* **1988,** *22*, 91–103.

19. Fenton, H. J. Oxidative Properties of the H_2O_2/Fe^{2+} System and its Application. *J. Chem. Soc.* **1884,** 65, 889–899.

20. Arnold, S. M.; William, J. H.; Robin, F. H. Degradation of Atrazine by Fenton's Reagent: Condition Optimization and Product Quantification. *Environ. Sci. Technol.* **1995,** *29*(8), 2083–2089.

21. Esplugas, S.; Marco, A.; Chamarro, E. Use of Fenton Reagent to Improve the Biodegradability of Effluents. *Proc. Int. Reg. Conf. Ozonat. AOPs Water Treat.* 1998, *23–25*.

22. Upelaar, G. F.; Meijers, R. T.; Hopman, R.; Kruithof, J. C.. Oxidation of Herbicides in Groundwater by the Fenton Process: A Realistic Alternative for O_3/H_2O_2 Treatment?. *Ozone Sci. Eng.* **2000,** *22*(6), 607–616.

23. Marina, T.; Veressinina, Y.; Munter, R. Advanced Oxidation Processes for Degradation of 2, 4-Dichlo-and 2, 4-Dimethylphenol. *J. Environ. Eng.* **1998,** *124*(8), 690–694.

24. Marina, T.; Anna Goi, A. Degradation of Nitrophenols with the Fenton Reagent. *Proc. Estonian Acad. Sci. Chem. Y.* **1999,** *48*(4), 163–173.

25. Zerva, C; Peschos, Z.; Poulopoulos, S. G.; Philippopoulos, C. J. Treatment of Industrial Oily Wastewaters by Wet Oxidation. *J. Hazard. Mater.* **2003,** *97*(1–3), 257–265.

26. Rivas, F. J., Kolaczkowski, S. T.; Beltrán, F. J.; McLurgh, D. B. Development of a Model for the Wet Air Oxidation of Phenol Based on a Free Radical Mechanism. *Chem. Eng. Sci.* **1998,** *53*(14), 2575–2586.

27. Kyoung-Hun, K.; Ihm, S. K. Heterogeneous Catalytic Wet Air Oxidation of Refractory Organic Pollutants in Industrial Wastewaters: A Review. *J. Hazard. Mater.* **2011,** *186*(1), 16–34.

28. Stüber, F.; Font, J.; Fortuny, A.; Bengoa, C.; Eftaxias, A.; Fabregat, A. Carbon Materials and Catalytic Wet Air Oxidation of Organic Pollutants in Wastewater. *Topics Catal.* **2005,** *33*(1–4), 3–50.

29. Perkow, H.; Steiner, R.; Vollmuller, H. Wet Air Oxidation-a Review. *Ger. Chem. Eng.* **1981,** *4*, 193–201.

30. Zhang, H.; Fei, C.; Zhang, D.; Tang, F. Degradation of 4-Nitrophenol in Aqueous Medium by Electro-Fenton Method. *J. Hazard. Mater.* **2007,** *145*(1–2), 227–232.

31. Wu, Jie, Zhang, H.; Oturan, N.; Wang, Y.; Chen, L.; Oturan, M. A. Application of Response Surface Methodology to the Removal of the Antibiotic Tetracycline by Electrochemical Process using Carbon-Felt Cathode and DSA (Ti/RuO$_2$-IrO$_2$) Anode. *Chemosphere* **2012,** *87*(6), 614–620.

32. Oturan, N.; Enric B.; Oturan, M. A. Unprecedented Total Mineralization of Atrazine and Cyanuric Acid by Anodic Oxidation and Electro-Fenton with a Boron-Doped Diamond Anode. *Environ. Chem. Lett.* **2012,** *10*(2), 165–170.

33. Commentary, Tech. Advanced Oxidation Processes for treatment of Industrial Wastewater. *An EPRI Commun. Environ. Cent. Publ.* **1996,** 1.

34. Prengle, H. W. Experimental Rate Constants and Reactor Considerations for the Destruction of Micropollutants and Trihalomethane Precursors by Ozone with Ultraviolet Radiation. *Environ. Sci. Technol.* **1983,** *17*(12), 743–747.

35. Agustina, T. E., Ang, H. M.; Vareek, V. K. A Review of Synergistic Effect of Photocatalysis and Ozonation on Wastewater Treatment. *J. Photochem. Photobiol. C. Photochem. Rev.* **2005,** *6*(4), 264–273.

36. Gurol, M. D.; Vatistas, R. Oxidation of Phenolic Compounds by Ozone and Ozone+ UV Radiation: A Comparative Study. *Water Res.* **1987,** *21*(8), 895–900.

37. Peyton, G. R.; Huang, F. Y.; Burleson, J. L.; Glaze, W. H. Destruction of Pollutants in Water with Ozone in Combination with Ultraviolet Radiation. 1. General Principles and Oxidation of Tetrachloroethylene. *Environ. Sci. Technol.* **1982,** *16*(8), 448–453.

38. Neyens, E.; Baeyens, J. A Review of Classic Fenton's Peroxidation as an Advanced Oxidation Technique. *J. Hazard. Mater.* **2003,** *98*(1-3), 33–50.

39. Alfano, O. M.; Rodolfo, J. B.; Alberto, E. C. Degradation Kinetics of 2, 4-D in Water Employing Hydrogen Peroxide and UV Radiation. *Chem. Eng. J.* **2001,** *82*(1–3), 209–218.

40. Gogate, P. R.; Pandit, A. B. A Review of Imperative Technologies for Wastewater Treatment II: Hybrid Methods. *Adv. Environ. Res.* **2004,** *8*(3–4), 553–597.
41. Litter, M. I. Introduction to Photochemical Advanced Oxidation Processes for Water Treatment. *Environmental Photochemistry Part II.* Springer: Berlin, Heidelberg, 2005; pp 325–366.
42. Nezamaddin, D.; Salari, D.; Khataee, A. R. Photocatalytic Degradation of Azo Dye Acid Red 14 in Water: Investigation of the Effect of Operational Parameters. *J. Photochem. Photobiol. A. Chem.* **2003,** *157*(1), 111–116.
43. Trapido, M.; Hirvonen, A.; Veressinina, Y.; Hentunen, J.; Munter, R. Ozonation, Ozone/UV and UV/H_2O_2 Degradation of Chlorophenols. *J. Int. Ozone Assoc.* **1997,** 75–96.
44. Nicole, I.; De Laat, J.; Dore, M. Evaluation of Reaction Rate Constants of OH Radicals with Organic Compounds in Diluted Aqueous Solutions using H_2O_2/UV Process. *Proc. 10th Ozone World Congr.* **1991,** 279–290.
45. Bischof, H.; Hofl, C.; Schonweitz, C.; Sigl, G.; Wimmer, B.; Wabner, D. UV–Activated Hydrogen Peroxide for Ground and Drinking Water Treatment–Development of Technical Process. *Proc. Reg. Conf. Ozone, UV-light, AOPs Water Treatm.* **1996.**
46. Commentary, Tech. Advanced Oxidation Processes for Treatment of Industrial Wastewater. *EPRI Commun. Environ. Cent. Publ.* **1996,** 1.
47. Trapido, M.; Veressinina, Y.; Kallas, J. Degradation of Nitrophenols by Ozone Combined with UV-Radiation and Hydrogen Peroxide. *Proc. Int. Conf. on Application of Ozone and also on UV and Related Ozone Technologies at Wasser Berlin;* 2000; pp 421–435
48. Lewis, N. M., Topudurti, K.; Foster, R. *A Field Evaluation of the UV/Oxidation Technology to Treat Contaminated Groundwater.* US Environmental Protection Agency, 1990.
49. Ruppert, Gerald, Bauer, R.; Heisler, G.; Novalic, S. Mineralization of Cyclic Organic Water Contaminants by the Photo-Fenton Reaction-Influence of Structure and Substituents. *Chemosphere* **1993,** *27*(8), 1339–1347.
50. Mahadwad, O. K.; Parikh, P. A.; Jasra, R. V.; Patil, C. Photocatalytic Degradation of Reactive Black-5 Dye using TiO2-Impregnated Activated Carbon. *Environ. Technol.* **2012,** *33*(3), 307–312.
51. Ollis, D. F.; Pelizzetti, E.; Serpone, N. Photocatalyzed Destruction of Water Contaminants. *Environ. Sci. Technol.* **1991,** *25*(9), 1522–1529.
52. Yoon, J. W.; Baek, M. H.; Hong, J. S.; Lee, C.H.; Suh, J. K. Photocatalytic Degradation of Azo Dye using TiO_2 Supported on Spherical Activated Carbon. *Korean J. Chem. Eng.* **2012,** *29*(12), 1722–1729.
53. Bailey, J.; Bemberis, I.; Presti, J. *Phase I Final Report-Shipboard Sewage Treatment System. General Dynamics Electric Boat Division*; NTIS, 1971.
54. Bemberis, I. Membrane Sewage Treatment Systems: Potential for Complete Wastewater Treatment. *Am. Soc. Agric. Eng.* **1971.**
55. Mutamim, N. S. A.; Noor, Z. Z.; Abu Hassan, M. A.; Olsson, G.. Application of Membrane Bioreactor Technology in Treating High Strength Industrial Wastewater: A Performance review. *Desalination* **2012,** *305*, 1–11.

56. Tan, J. M., Guanglei, Q.; Ting, Y. P. Osmotic Membrane Bioreactor for Municipal Wastewater Treatment and the Effects of Silver Nanoparticles on System Performance. *J. Clean. Prod.* **2015,** *88,* 146–151.

57. Rodríguez, F. A.; Poyatos, J. M.; Reboleiro-Rivas, P.; Osorio, F., González-López, J., Hontoria, E.; Kinetic Study and Oxygen Transfer Efficiency Evaluation using Respirometric Methods in a Submerged Membrane Bioreactor using Pure Oxygen to Supply the Aerobic Conditions. *Bioresour. Technol.* **2011,** *102*(10), 6013–6018.

CHAPTER 5

Green Energy: Renewable Power Generation from Solar PV Cells

Impact of Irradiation, Materials and Aging on Solar Power Generation, and Green Materials for Solar Power Extraction

M. A. ASHA RANI[1*], M. CHAKKARAPANI[2], and PRANJIT BARMAN[3]

[1]*Department of Electrical Engineering, National Institute of Technology, Silchar 788010, Assam, India*

[2]*Department of Electrical and Electronics Engineering, Madanapalle Institute of Technology and Science, Madanapalle 517325, Andhra Pradesh, India*

[3]*Department of Chemistry, National Institute of Technology, Silchar 788010, Assam, India*

*Corresponding author.
E-mail: asharani@ee.nits.ac.in; asha_siju@yahoo.co.in*

ABSTRACT

Concern over the limited stock of conventional energy sources such as coal and other petroleum products has fuelled efforts toward the development of renewable sources of energy that have a lesser footprint on the environment. Materials and technologies play a vital role that can offer promising solutions to achieve renewable and sustainable pathways for the future. Of these renewable sources, solar radiation is the most abundant and freely available and can be directly harnessed by the use of photovoltaic (PV) modules. This chapter is based on the renewable power generation from solar PV cells. In this chapter, we will discuss about the renewable power generation scenario; working of a solar cell; model of solar PV module; the impact of PV cell material in PV characteristics; the effect of DC link capacitor material in power converters; challenges involved in solar power generation, and their mitigation techniques; and green materials using green chemistry for fabricating the future solar PV cell and DC link capacitors.

5.1 INTRODUCTION

Escalating demand for energy, the depleting fossil fuel reserves and the growing concerns over environmental degradation have brought in new paradigms in the global perspective and in India as well. Concern for clean energy reducing the dependence on fossil fuels has provided a huge impetus to the search for alternative renewable energy sources that have a lower carbon footprint on the environment. The use of renewable energy supplements the energy needs and thereby reduces the environmental impact, which is the prime concern of many countries around the world. Renewable energy sources such as wind and solar become more competitive in the market owing to the technological advancements, reduced cost, and governmental incentives. Among these renewable alternatives, solar energy is freely available in unlimited quantity, and the conversion of solar energy to electricity using photovoltaic (PV) modules does not require any moving parts. The amount of solar radiation striking our earth's surface is about 10,000-times higher than the current global electrical energy consumption. PV cells convert sunlight directly to electricity and can be influential in meeting the world's energy demand. Hence, this technology is gaining popularity due to its reducing cost, substantial investment in research and development, and attractive government subsidies.

5.2 RENEWABLE POWER GENERATION SCENARIO

Based on the geographical location and climatic conditions, there is a tremendous scope of harnessing the solar power using the unutilized space and wastelands around the fields and buildings. Thus, installing and commissioning solar power plants will be the appropriate cost effective solution for an environment friendly power generation mitigating the dependence of fossil fuels.

Solar energy has a tremendous potential with an average availability of 300 solar days per year. Via several energy policies, state and central governments of various countries are providing incentives for solar power generation and also for various solar applications in addition to the technical forums and seminars/conferences conducting worldwide for creating awareness.

Confederation of Indian Industry (CII) generally organizes conferences in India on an annual basis at different parts of the country, which serves as an ideal platform for participants to gain a strong understanding of the power scenario in the near future, challenges to be addressed, opportunities for industries and the role of various stakeholders in this. The prime objectives of the conference are to explore the opportunities for organizations in the power and renewable sector; to connect with key stakeholders on the issues pertaining to power and renewable sector, and understand the measures to be taken at various levels that will ensure reliable power for all, both in quality and quantity; to convene thoughts and ideas on making power sector in South India more vibrant to support the national GDP goals; to bring in experts from across the globe to know the clean energy drive; and to understand new and innovative technologies, systems, and business models in power sector.

Similarly, Ministry of Science and Technology under Government of India regularly conducts seminars and is looking for innovative ideas in order to overcome the challenges related to global climate change and energy crisis.

In India, government has set a goal that solar energy should contribute to 8% of India's total consumption of energy by 2022 such that solar is going to play a key role in shaping the future of Indian power sector.

Also, Government of India has taken initiatives to enhance renewable power generation in an effective way that includes no further installation of thermal power plants to avoid carbon emission, electrifying all the light motor vehicles by 2030, and electrifying all the vehicles including heavy duty vehicles by 2050 to minimize the emission of CO_2 by 0.3 kg/unit of thermal power generated.

5.3   WORKING OF SOLAR CELL

A solar cell is a photo detector with a p–n junction illuminated to generate DC current. A typical silicon solar cell is composed of a thin wafer layer of n-type (pentavalent dopant) silicon as top layer, and a thicker p-type (trivalent dopant) silicon layer at bottom as shown in Figure 5.1. The contact between the two materials forms the p–n junction, which has a built-in electrical field in the depletion region.

The photovoltaic process of current generation in a solar cell involves two steps. The first step is the absorption of incident photons to generate electron–hole pairs. Electron–hole pairs can be created only if the incident photons have energy greater than the semiconductor bandgap. In the second step, electrons and holes are separated by the electrical field in the junction depletion region and flows through the external circuit.

Generally, solar panels are connected in series or parallel to meet the energy demand as detailed below.

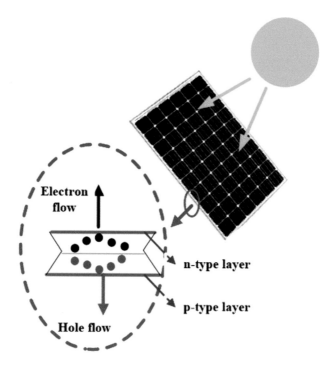

FIGURE 5.1   Typical solar cell structure.

(a) Connecting Solar Panels in Series

Generally, solar panels are connected in series to increase the terminal voltage of the solar PV system. By connecting the solar panels in series (similar solar panels), the total output voltage of the array obtained will be the sum of individual panel voltages as shown in Figure 5.2.

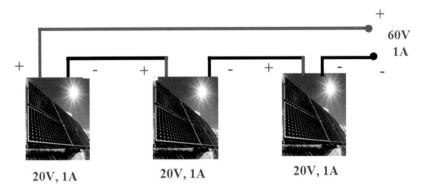

FIGURE 5.2   Series connection of solar panels.

(b) Connecting Solar Panels in Parallel

Solar panels are connected in parallel to enhance the overall system current. By connecting the solar panels in parallel, the total output voltage remains the same as if in a single panel, whereas the output current will be the sum of the output currents of individual panels as shown in Figure 5.3.

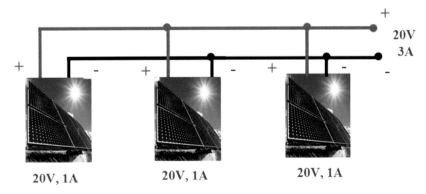

FIGURE 5.3   Parallel connection of solar panels.

5.4   MODEL OF A SOLAR PV MODULE

A solar PV module consists of several cells in series. A single cell, expressed by its single diode equivalent circuit, is shown in Figure 5.4. Therefore, the characteristics of the PV module can be obtained by connecting the models of several such cells in series. The relation between voltage and current of a PV module can be expressed as:

$$I_{PV} = I_{ph} - I_s \left[\exp\left(\frac{V_{PV} + R_s I_{PV}}{A} \right) - 1 \right] - \left[\frac{V_{PV} + R_s I_{PV}}{R_{sh}} \right] \quad (5.1)$$

where, I_{PV} is the current generated by the module; I_{ph} is the light generated current; I_s is the reverse saturation current; V_{PV} is the PV voltage; and $A = n_s kT/q$. Here, n_s is the number of series-connected cells; k is the Boltzmann's constant; T is the temperature of the module in Kelvin; q is the elementary charge; and R_s and R_{sh} are the series and shunt resistances, respectively. The light generated current (Kadri et al. (2011)) is given by the equation:

$$I_{ph} = I_{sco} \left(\frac{G}{G_o} \right) (1 + \alpha_1 (T - T_o)) \frac{(R_s + R_{sh})}{R_{sh}} \quad (5.2)$$

where, I_{sco} is the short circuit current of the module at standard irradiance G_o (1000 W/m^2) and standard temperature T_o (25°C), and α_1 is the module's temperature coefficient for current.

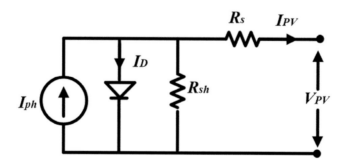

FIGURE 5.4   Single-diode equivalent circuit of a PV cell.

5.5   IMPACT OF SOLAR IRRADIATION ON PV CHARACTERISTICS

From eqs (5.1) and (5.2), it is clear that the module output current is directly proportional to the irradiance level G. The specifications of a typical PV module at standard test conditions (STC) of 1000 W/m^2, 25°C and air mass (AM) are given in Table 5.1.

TABLE 5.1   PV Specifications at STC 1000 W/m^2, 25°C, AM 1.5.

PV power	80 W
Open circuit voltage, V_{oc}	22 V
Short circuit current	4.7 A
MPP voltage	18 V
MPP current	4.44 A

Figures 5.5 and 5.6 illustrate the current–voltage characteristics and the power–voltage characteristics of the above PV module.

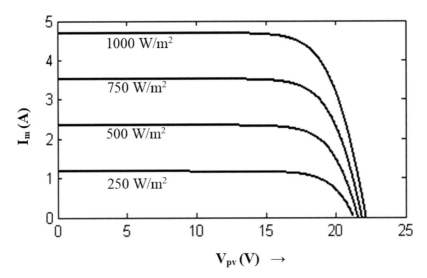

FIGURE 5.5   *I–V* characteristics of the PV module for different irradiance levels.

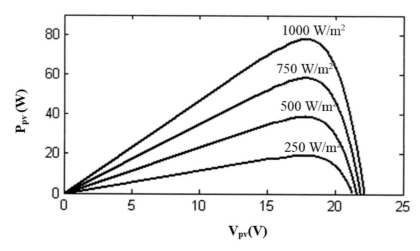

FIGURE 5.6 $P-V$ characteristics of the PV module corresponding to the $I-V$ curves of Figure 5.2.

It is observed from Figure 5.6 that there is a unique operating point called the maximum power point (MPP), at which the PV module generates power that is greater than that at all other points on the curve. Further, the maximum power and the voltage at which the maximum power occurs are different under different insolation conditions as shown in Figure 5.6. Therefore, it is clear that the MPP of the PV module varies according to the solar irradiance, under varying environmental conditions.

Here, the variation in short circuit current (I_{sc}) is proportional to the solar irradiation, whereas open circuit voltage (V_{oc}) varies logarithmically with irradiation and hence the power also changes with respect to irradiation. From the characteristics (Fig. 5.6), it is evident that Vmpp is decreasing when the irradiation decreases from 1000 to 250 W/m². However, Vmpp is almost constant with slight decrement with respect to the irradiation.

In addition, the important thing to be noted is that there is a significant impact of the material used for making the PV cell on the $P-V$ characteristics of the module since all the parameters mentioned above such as PV power, open circuit voltage, short circuit current, MPP voltage, and MPP current will vary due to the variation of series and shunt resistances, R_s and R_{sh}, respectively, and α_1 is the module's temperature coefficient for current in eq 5.2. This is discussed in Section 5.4.

5.6 IMPACT OF PHOTOVOLTAIC CELL MATERIAL IN PV CHARACTERISTICS

The impact of photovoltaic cell material in PV characteristic is discussed in this section by considering a PV system with resistive load as shown in Figure 5.7.

The set-up consists of a PV array followed by a DC–DC boost converter feeding a resistive load. The boost converter is employed for MPPT operation as it presents a variable resistance to the PV array. In addition, the voltage generated by the PV string is not sufficient to feed the load directly. Here, boost converter plays the major role to increase the voltage to the required level.

The operation of boost converter for MPPT is as discussed below. The input–output relations of the voltage, current, and resistance can be written as

$$V_0 = \frac{V_{PV}}{1-D} \tag{5.3}$$

$$I_0 = I_{PV}\left(1-D\right) \tag{5.4}$$

$$R_{PV} = R_o\left(1-D\right)^2 \tag{5.5}$$

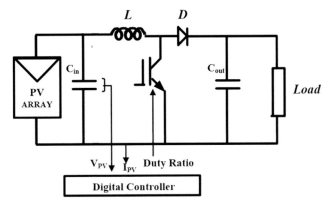

FIGURE 5.7   Schematic diagram of the PV system feeding a resistive load through a PV array and boost converter.

Here, R_{pv} and R_o are the input and output resistance of the PV panel and D is the duty ratio of the boost converter.

When a constant resistance is connected at the output of the converter, the input resistance (on an average) is dependent on the duty ratio D of the converter as given by eq 5.5. By varying the duty ratio, the PV module can be made to operate at different voltages, and the operating point is given by the point of intersection of the load line and the I–V characteristic as depicted in Figure 5.8.

Thus, as the load line varies from R_1 to R_4 by varying the duty ratio from D_1 to D_4, various operating points on the I–V curve can be accessed. Out of these points, one of the points will be the MPP, corresponding to a load of R_{mpp}. Thus, during MPPT, duty ratio is varied such that the reflected load resistance on the PV array is R_{mpp}.

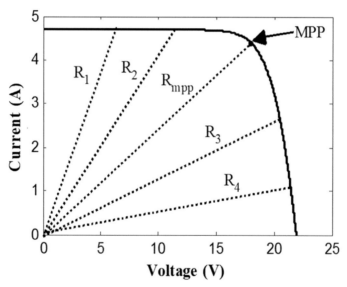

FIGURE 5.8   Variation in PV operating point with reflected load resistance.

Moreover, from Figure 5.6 it is clear that the series and shunt resistances, R_s and R_{sh}, respectively, and the module's temperature coefficient for current will vary based on the irradiation as well as the material of the PV cell; and hence, the PV power, open circuit voltage, short circuit current, MPP voltage, and MPP current. Similar is the case with R_{mpp}, the reflected load resistance on the PV array at MPP.

5.6.1 CLASSIFICATION OF PV MODULES BASED ON MATERIAL USED

Different types of PV modules available in the market and its classification based on the material used is discussed in this section. Solar PV modules currently used in PV power plants are of crystalline silicon and thin film modules (Fig. 5.9). Based on the process of growth of silicon crystal, crystalline modules are further classified as monocrystalline and polycrystalline modules. Whereas, thin films modules are classified based on the compounds used in fabricating the modules.

(A) Crystalline Silicon (c-Si) Modules

c-Si modules are made of solar cells encapsulated between a transparent front glass and a backing material made of plastic or glass. As mentioned above, crystalline modules are further classified as monocrystalline and polycrystalline modules. The monocrystalline silicon wafers are made through Chocklarsky process by slicing a large single crystal ingot. Whereas, polycrystalline silicon wafers are made by slicing the cast molten multi-silicon ingots and are larger than mono-crystalline wafers. Polycrystalline cells are cheaper than monocrystalline cells, but are less efficient than monocrystalline ones.

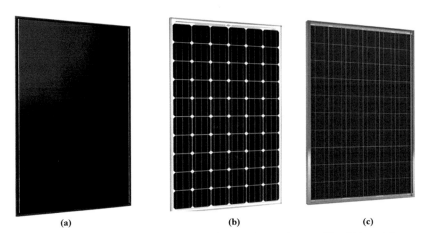

(a) (b) (c)

FIGURE 5.9 (a) Thin film, (b) monocrystalline, and (c) polycrystalline PV modules.

(B) Thin film Modules

Thin film modules are broadly classified as amorphous silicon, cadmium telluride, copper indium (gallium) di-selenide, and heterojunction with intrinsic thin film layer (HIT) as discussed below.

(a) Amorphous Silicon (a-Si)

In a-Si module, atoms form a continuous random network unlike as in c-Si. a-Si modules are thinner and are available at low cost than c-Si owing to the higher efficiency of a-Si to absorb light than c-Si's, and also due to the possibility of depositing on both rigid and flexible low-cost substrates. Thus, facilitating a-Si as an ideal choice for different applications where the concern is of low cost rather than efficiency.

(b) Cadmium Telluride (CdTe)

CdTe modules are derived from cadmium and tellurium. This module is fabricated by depositing a semiconductor film stack on a transparent conducting oxide-coated glass. These modules are capable of producing high energy output across a wide range of climatic conditions.

(c) Copper Indium (Gallium) Di-Selenide (CIGS/CIS)

CIGS/CIS is a semiconductor module derived from copper, indium, gallium, and selenium with a capability of capturing more light than c-Si. However, these modules need thicker films than a-Si PV modules. Also, these modules are capable of offering the highest conversion efficiency among all the thin film PV module technologies.

(d) Heterojunction with Intrinsic Thin Film Layer (HIT)

Heterojunction with intrinsic thin film layer solar cell is composed of ultrathin amorphous silicon layers as the surrounding material for a mono-thin-crystalline silicon wafer. Even though HIT modules are more expensive than crystalline modules, their reliability and efficiency are far more than crystalline modules.

A comparison between different types of PV modules is as furnished in Table 5.2 shown below. Crystalline silicon dominates the solar market globally because it has the highest energy efficiency and longer lifespan compared to other PV modules.

TABLE 5.2   Comparison of PV Modules.

Type of PV module	Sensitivity to temperature	Efficiency of current (%)
Monocrystalline	√	15–22
Polycrystalline	√	14–17
Thin film	x	5–13
HIT films	x	20–22

5.7   IMPACT OF AGING OF PV MODULES (DEGRADATION) IN RENEWABLEPOWER GENERATION

In addition to the high cost of PV panels, yet another challenge in the solar power sector is the aging of the PV modules leading to the degradation of the photovoltaic (PV) module, which in turn reduces the output power, owing to the extreme environmental operating conditions. Due to degradation, there is a considerable change in the series and shunt resistance of the PV module leading to reduced output power. PV modules experience fast degradation than expected since, the PV modules are operated for a long time in the outdoor conditions without proper maintenance and due to its corrosive nature.[1] However, the degradation rate depends upon the operating conditions.

Considering the diverse climatic conditions, the PV modules exposed to hot and dry climatic zones experience higher degradation rate that is 1.55% per year.[2] The degradation of PV panel can be categorized as optical degradation, cell degradation, degradation of cell/module interconnects, light induced degradation, temperature-induced degradation, and potential induced degradation.[3] Owing to the degradation in the PV module, there will be considerable changes in the shunt resistance (R_{sh}), series resistance (R_{se}), voltage at maximum power point (V_{mpp}), current at maximum power point (I_{mpp}), and fill factor leading to severe power loss in the PV system. Degradation can lead to either increase in R_{se} or decrease in R_{sh}.

5.7.1   EFFECT OF R_{SH} AND R_{SE} DEGRADATION ON V_{MPP}

Due to degradation, as R_{sh} decreases from its nominal value with a constant irradiation for example, at 1000 W/m² at standard test conditions (STC), V_{mpp} of the panel increases and vice versa. Similarly, as the series resistance R_{se} increases from its nominal value at STC, the V_{mpp} of the panel also decreases and vice versa.

5.8   EFFECT OF DC LINK CAPACITOR MATERIAL IN POWER CONVERTERS/SOLAR INVERTER

Inverters are inevitable for interfacing PV modules to the utility grid, which plays key role in the efficiency, cost, lifetime, and stability of PV systems. The performance of PV inverters mainly depends on the power electronic devices used in the power converters. Si devices are widely available in market in all forms of semiconductor devices with a maximum junction temperature limit of 150°C. Hence, while designing any semiconductor device using Si, the temperature of the device should be below the specified limit. Moreover, fabricating higher voltage rating Si MOSFETs is not feasible due to the high cost as the device requires a large silicon die area as the breakdown voltage increases. In addition, the switching frequency of Si devices is also limited due to the heat generated by the devices. Due to the above-mentioned limitations of Si devices, now people are looking for power electronic devices with reduced size, weight, cost, power density, and improved efficiency close to cent percentage. Hence, there is an increased attention toward other wide bandgap semiconductors such as silicon carbide (SiC), Gallium arsenide (GaAs), Gallium nitride (GaN), Diamond, etc. A few properties of these wide bandgap semiconductor materials are given in Table 5.3.[4]

TABLE 5.3   Properties of Wide Bandgap Semiconductor Materials.

Property	Si	GaAs	SiC	GaN	Diamond
Bandgap, E_g(eV)	1.12	1.43	3.26	3.45	5.45
Dielectric constant	11.9	13.1	10.1	9	5.5
Electric breakdown field, E_c(kV/cm)	300	400	2200	2000	10,000
Electron mobility (cm²/Vs)	1500	8500	1000	1250	2200
Hole mobility (cm²/Vs)	600	400	115	850	850
Thermal conductivity (W/cm.K)	1.5	0.46	4.9	1.3	22
Saturated electron	1	1	2	2.2	2.7

It is clear from the table that, diamond is the best choice among these materials. However, cost is the barrier in designing semiconductors with diamond. Finally, considering the practical difficulties, now manufacturers have chosen silicon carbide (SiC) in place of Si devices. However, GaN is also a possible solution.

SiC-based devices were commercialized by CREE Inc. in 1987 with SiC-based Schottky diodes with a voltage blocking capability of 600 V. Whereas, the voltage blocking capability of Si devices is limited to 300 V. Nowadays, SiC semiconductor devices are being designed for high-temperature, high-power, and high-radiation conditions as well unlike Si-based semiconductors. This invention is making revolutionary changes in the semiconductor design field since SiC devices are suitable for high voltage and temperature ranges with superior switching characteristics than silicon-based devices.

5.9 CHALLENGES INVOLVED IN SOLAR POWER GENERATION

PV sources inherently have a unique operating point called the maximum power point (MPP) at which maximum power may be extracted from them. However, this point is never constant with respect to time; it changes with solar irradiance, temperature, and load profile.[5] In order to extract the maximum possible power from the PV system under varying conditions, a maximum power point tracking (MPPT) scheme is generally adopted[6], which requires power electronic interfaces such as DC/DC converters or DC/AC inverters.

Depending on the current and voltage required by the system load, series–parallel connections of individual PV modules are usually adopted. A "string" refers to a set of series-connected PV modules. Under normal operating conditions, each module in a string receives the same amount of irradiance, and the PV characteristics of the string resembles that of the single module, with the voltage increased by as many times as the number of series modules and the current being the same. In such cases, the P–V curve of the string exhibits a single peak.

However, as PV modules are installed in outdoor environments, it is practically possible that the modules receive unequal levels of irradiance, and this phenomenon is termed as partial shading. Partial shading is usually caused by the shadow of buildings or other structures, trees, clouds, or dirt.[7] Under partial shading conditions, the hot spot problem occurs wherein the shaded module becomes reverse biased and hence behaves as a load. To eliminate this problem, bypass diodes are connected in each module to safely bypass the reverse current. Due to the presence of these diodes, the string $P-V$ characteristic is drastically affected on the occurrence of partial shading.[8] Under partial shading conditions, multiple peaks are developed, with several local MPPs (LMPPs), and the location of the global MPP (GMPP) depends on the degree of mismatch in the modules' irradiance. Conventional MPPT algorithms are not designed to handle local peaks efficiently. When the MPPT algorithm is locked on a local MPP, the PV system efficiency falls dramatically.

In addition to partial shading, PV systems are subjected to various failures or faults among the PV arrays, power conditioning units, batteries, wiring, and utility interconnections.[9] It is difficult to shut down PV modules completely during arrays faults, since they are energized by sunlight in the daytime. In general, PV array faults can be classified as line-ground faults, line–line faults and open–circuit faults. Among these faults, line–line faults and ground faults are the most common faults in solar PV arrays, which potentially involve large fault currents. Without proper fault detection or protection, they could cause severe problems, such as dc arcs and even fire hazards.[10] For example, a multipoint ground fault reportedly[14] caused a fire in a PV power plant in Bakersfield, California and in another instance; a fire was caused by a double-ground fault in a large PV power plant in Mount Holly, North California.[15] These fire incidents not only underscore the weaknesses in conventional fault detection and protection schemes for PV arrays, but also reveal the urgent need for a better protection system. Among these two abnormal conditions, viz. partial shading and faults, the former will exist for a short period of time, whereas the latter would persist over time. In a PV installation, there is a need to differentiate partial shading from faulty conditions in order to avoid inadvertent shutdowns.

5.10 MITIGATION TECHNIQUES FOR OVERCOMING THE CHALLENGES INVOLVED IN SOLAR POWER GENERATION

Several efficient control strategies are reported to detect partial shading and faults in PV array for the enhanced extraction of PV power output to improve the system efficiency and reliability.

Without proper fault detection and protection, they could cause severe problems in the PV array, such as dc arcs and even fire hazards[10]. Conventional protection systems for PV arrays consist of overcurrent protection devices (OCPD), Arc-fault circuit interrupters (AFCI) and ground fault protection devices (GFPD) as stated in US NEC[16]. The challenges involved in the protection of PV arrays have been discussed in[17], including the negative impact of environmental conditions, MPPT, and blocking diodes in fault detection, along with the inability of OCPD to detect faults under certain conditions.

Faults are detected by comparing the actual electrical array quantities with expected array quantities.[18] A similar approach is adopted in[19], but the measured considered here are the irradiances in the horizontal plane and in the PV module plane, the ambient temperature, as well as electrical quantities at the DC and AC side of the PV system, which are fed into a simulation to calculate the normalized capture losses. A fault detection method has been proposed[20], in which all PV string currents are measured and tested under outlier detection rules. Even the best performing outlier detection rule for short-circuit faults produces false alarms during normal conditions before and after the occurrence of partial shading. An overview of different fault detection schemes and recommendations for improvement are reported in[21]. A method to detect faults and partial shading using the measured array voltage, current, and irradiance is proposed[22] for operation under normal operating condition, partial shading, and faults.

5.11   GREEN CHEMISTRY FOR GREEN ENERGY

Owing to the increasing demand of electric power, were materials plays a major role in sustainable power extraction, green chemistry opens new doors for a green future. There is a symbiotic relationship between green chemistry and renewable energy. The usage of unsafe substances or materials in the design, synthesis, and application of chemical products brings new environmental problems like global warming, acid rain, ozone layer depletion, and many other harmful side effects, which necessitates that there is requirement for practicing green chemistry and materials. This facilitates, minimizing the consumption of materials and energy, produces least or zero waste materials, economically strong and safe, making it the best solution.

There are twelve basic principles in green chemistry: (1) prevent waste, (2) maximize atom economy, (2) less hazardous chemical synthesis, (4) safer chemicals and products, (5) safer solvents and reaction conditions, (6) increase energy efficiency, (7) use renewable feed stocks, (8) avoid chemical derivatives, (9) use catalysts, (10) design chemicals and products to degrade after use, (11) analyze in real time to prevent pollution, and (12) minimize potential for accidents.[11–13] Applying a few of these principles during synthesis of products is known as green chemistry approach. The next section of this chapter discusses about the scope of using green chemistry for fabricating solar cells with the help of organic materials.

5.11.1   GREEN CHEMISTRY FOR ORGANIC SOLAR CELLS

(a) Conducting Polymer

The strategies for producing conjugated polymers using green chemistry are discussed in this section. Polymers are usually used as insulators. However, in the mid 1970s, polyacetylene was accidentally fabricated by the scientist Shirakawa, the first polymer capable of conducting electricity. It is an organic polymer with the repeating unit $(C_2H_2)_n$. Later, in 2000, the chemistry Nobel prize was awarded to Alan J. Heeger, Alan G MacDiarmid and Hideki Shirakawa for the discovery and study of conducting polymers.

Usually the synthesized conducting polymers exhibit very low conductivities. However, conductivity can be increased by adding dopants (p-type dopants or electron acceptor and n-type dopant or electron donar dopants). Synthetic or conducting polymers are used as conductors over metals owing to the following advantages like lighter in weight, easy, and less energy consumption during processing, corrosion resistant, cheapest materials, easy transportation, easy to handle, etc. Due to this capability, they are used in the fabrication of electronic devices, solar energy conversion, rechargeable batteries, sensors, etc.

For a polymer to behave as a conducting polymer, the polymer chain should contain pi-electrons/lone pair of electrons or vacant p-orbitals. Few examples of conducting polymers that can be used for fabricating solar cells are given in Table 5.4.

TABLE 5.4   List of Conducting Polymers, their Energy Gap and Conductivity.

Polymer	Discovery	Structure	Energy bandgap (eV)	Conductivity (S/cm)
Polythiophene	1981		2.1	$10-10^3$
Polyacetylene	1977		1.5	$10^3-1.7 \times 10^5$
Polyaniline	1980		3.2	$30-200$
Polypyrrole	1979		3.1	$10^2-7.5 \times 10^3$
Poly (3,4-ethylene-dioxythiophene)	1980		1.1	300
Poly (p-phenylene vinylene)	1979		2.5	$3-5 \times 10^3$
Polyphenylene and Polyparaphenylene	1979		3.0	10^2-10^3

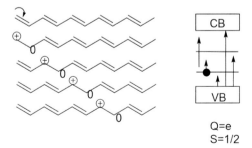

FIGURE 5.10   Mechanism of conduction in polymers and electron transfer from valence band (VB) to conduction band (CB).

KEYWORDS

- renewable power generation
- solar photovoltaic (PV) module
- solar PV module modeling
- aging of PV modules
- wide bandgap semiconductor materials
- challenges in solar power generation
- organic solar cells

REFERENCES

1. Munoz, M.; Alonso-Garca, M.; Vela, N.; Chenlo, F. Early Degradation of Silicon pv Modules and Guaranty Conditions. *Solar Energy* **2011**, *85*(9), 2264–2274.

2. Chattopadhyay, S.; Dubey, R.; Kuthanazhi, V.; John, J. J.; Solanki, C. S.; Kottantharayil, A.; Arora, B. M.; Narasimhan, K. L.; Vasi, J.; Bora, B.; Singh, Y. K.; Sastry, O. S. All India Survey of Photovoltaic Module Degradation 2014: Survey Methodology and Statistics. In *Photovoltaic Specialist Conference (PVSC)*, 2015 IEEE 42nd, pp. 1–6, 2015.

3. Meyer, E.L.; van Dyk, E. E. Assessing the Reliability and Degradation of Photovoltaic Module Performance Parameters. *IEEE Trans. Reliab.* **2004**, *53*(1), 83–92.

4. Hudgins, J. L., Simin, G. S., Santi, E., Khan, M. A. An Assessment of Wide Bandgap Semiconductors for Power Devices. *IEEE Trans. Power Electron.* **2003**, *18*(3), 907–914.

5. Aureliano, M.; Galotto, L.; Sampaio, L. P.; Melo, G. D. A.; Canesin C. A. Evaluation of the Main MPPT Techniques for Photovoltaic Applications. *IEEE Trans. Ind. Electron.* **2013**, *60*, 1156–1167.

6. Jain, S.; Agarwal, V. A Single-Stage Grid Connected Inverter Topology for Solar PV System with Maximum Power Point Tracking. *IEEE Trans. Power Electron.* **2007**, *22*, 1928–1940.

7. Patel, H.; Agarwal, V. MATLAB Based Modeling to Study the Effects of Partial Shading on PV Array Characteristics. *IEEE Trans. Energy Convers.* **2018**, *23*, 302–310.

8. Balasubramanian, I. R.; Ganesan, S. I.; Chilakapati, N. Impact of Partial shading on the Output Power of PV Systems Under Partial Shading Conditions. *IET Power Electron.* **2014**, *7*, 657–666.

9. Harb, S.; Balog, R. Reliability of Candidate Photovoltaic Module Integrated-Inverter (PV-MII) Topologies–A Usage Model Approach. *IEEE Trans. Power Electron.* **2013**, *28*, 3019–3027.

10. Flicker, J.; Johnson, J. Analysis of Fuses for Blind Spot Ground Fault Detection in Photovoltaic Power Systems. *Solar America Board for Codes and Standards Report.* 2013.

11. Beyond Benign. Twelve Principles of Green Chemistry Manual, © 2010 Beyond Benign – A warner babcock foundation.

12. Anastas, P.; Warner, J. *Green Chemistry: Theory and Practice, United States Environmental Protection Agency (EPA)*; Oxford University Press: New York, 1998.

13. Anastas, P.; Warner, J. *Green Chemistry Pocket Guide, American Chemical Society*; Oxford University Press: New York, 1998.

14. Brooks, B. The bakersfield fire–a lesson in ground-fault protection, Solar Pro Magazine. 2011, 62–70.

15. NC PV DG Program SEPA presentation, Duke Energy, 2011, pp. 1–14.

16. Article 690-Solar Photovoltaic Systems, US National Electrical Code, 2011.

17. Zhao, Y.; De Palma, J. F.; Mosesian, J.; Lyons, R.; Lehman, B. Line–Line Fault Analysis and Protection Challenges in Solar Photovoltaic Arrays. *IEEE Trans. Ind. Electron.* **2013**, *60*, 3784–3795.

18. Stellbogen, D. Use of PV Circuit Simulation for Fault Detection in PV Array Fields. *IEEE Int. Conf. Photov.* Spec. **1993**, 1302–1307.

19. Silverstre, S.; Chouder, A.; Karatepe, E. Automatic Fault Detection in Grid Connected PV Systems. *Sol. Energy.* **2013**, *94*, 119–127.

20. Zhao, Y.; Lehman, B.; Ball, R.; Mosessian, J.; De Palma, J. Outlier Detection Rules for Fault Detection in Solar Photovoltaic Arrays. *IEEE Int. Conf. Appl. Power Electron.* Conf. Expo (APEC). **2013**, 2913–2920.

21. Alam, K.; Khan, F.; Johnson, J.; Flicker, J. A Comprehensive Review of Catastrophic Faults in PV Arrays: Types, Detection, and Mitigation Techniques. *IEEE J. Photov.* **2015**, *5*, 982–997.

22. Hariharan, R.; Chakkarapani, M.; Ilango, G. S.; Nagamani C. A. Method to Detect Photovoltaic Array Faults and Partial Shading in PV Systems. *IEEE J. Photov.* **2016**, *6*(5), 1278–1285.

Green Synthesized Carbon-Based Nanomaterials: Applications and Future Developments

ANU ROSE CHACKO, NEENA JOHN PLATHANAM, BINILA K KORAH, THOMAS ABRAHAM, and BEENA MATHEW*

School of Chemical Sciences, Mahatma Gandhi University, Kottayam, India

Corresponding author. E-mail: beenamscs@gmail.com

ABSTRACT

Carbon-based nanomaterials are the most valuable materials used in the modern field due to its high potential of application in almost all areas of living. Increase in crisis for energy and environmental degradation are the foremost challenges of using nanotechnology for sustainable development. Here lies the importance of green nanotechnology, which focuses on the use of green natural precursors for the development of eco-friendly processes and products. In this scenario, researchers have started the use of renewable, inexpensive, and abundant carbon-based nanomaterials for sustainable development. Even though there are enormous applications of carbon-based nanomaterials in various fields, this chapter proceeds with the applications of green synthesized nanomaterials for future perspectives. The major applications of carbon-based nanomaterials outlined involve prevention of environmental degradation, improvement of public health, energy efficiency, optimization, and industrial development. Green synthesized carbon nanomaterials such as carbon dots, carbon nanotubes (CNTs), graphene, fullerenes, and nanodiamonds are given special attention due to their excellent and efficient properties. This chapter provides the reader the current progress, highlighting the application in environment and energy-related fields of the carbon-based nanomaterials synthesized using green protocol.

6.1 INTRODUCTION

Carbon-based nanomaterials with its accountable properties and enormous applications have become an unavoidable part of development. They play a chief role in the prevention of environmental degradation, improvement of public health, energy efficiency optimization, wastewater reuse, and pollutant transformation. But the fossil fuel-based precursors used for the synthesis of carbon nanomaterials are a major challenge in light of the increase in global demand for energy. In order to address these issues, there is a high need for innovative and efficient sustainable solutions. Production of nanomaterials that could solve environmental problems without causing any harm to the environment or living organisms are considered as the essentialities for safer green nanotechnology. Nontoxic precursors which are renewable, eco-friendly, and inexpensive are excellent alternatives for traditional chemical sources.[1,2]

In this chapter, efforts are done to emphasize the applications of carbon-based nanomaterials that are synthesized from natural green precursors. And also highlight the fact that these green synthesized carbon nanomaterials could give better outcome compared to fossil fuel-derived carbon nanomaterials. This chapter proceeds with a main focus on the specialties and applications of green synthesized carbon nanomaterials such as carbon dots, CNTs, graphene, graphene oxide (GO), fullerene, and nanodiamond. Figure 6.1 illustrates all the potential applications of the green synthesized carbon-based nanomaterials. We expect that this will provide a brief compilation of all the applications and trends in the use of green synthesized carbon nanomaterials.

6.2 CARBON QUANTUM DOTS (CDs)

Fluorescent CDs are a new class of carbon nanoparticles that have recently emerged and have gained much interest as competitors of traditional semiconductor quantum dots (QDs). The observable superior properties of CDs include water solubility, low toxicity, biocompatibility, small size, fluorescence, and ease of modification. Zero-dimensional, size of <10 nm and quasi-spherical CDs have attracted much attention since 2006, as an environmentally friendly substitute for toxic QDs.[3] For many years, semiconductor QDs have been extensively investigated for their durable and tunable fluorescence emission properties, enabling their applications in sensing and bioimaging. However, semiconductor QDs have some limitations such as high toxicity due to the use of heavy metals in their production. Heavy metals are known to be very toxic even at relatively low doses, which would preclude any clinical study.[4,5] This was solved since 2004, with the invention of CDs having similar fluorescence properties.[6]

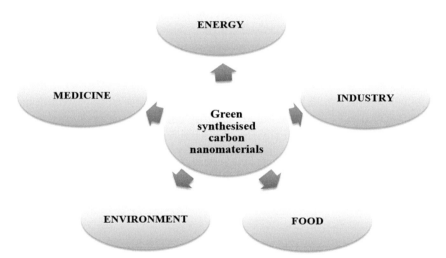

FIGURE 6.1 Schematic illustration of applications of green synthesized carbon nanomaterials.

The strategies for the synthesis of CDs are based on cutting larger carbon materials (top-down) or fusing smaller precursor molecules (bottom-up).[3,4] Generally, the bottom-up method has a high yield and it is convenient to introduce heteroatom doping in the synthesis process. Various methods of preparing CDs have been developed and most studies attempt to produce high-quality CDs using simple, cost-effective, size-controlled, or large-scale synthetic methods.[7] The preparation of organic CDs usually adopts the bottom-up approach including hydrothermal,[8] microwave treatment,[9] pyrolysis,[10] extraction,[11] etc. Among them, hydrothermal treatment is the most commonly used method. Organic carbon sources are eco-friendly natural products compared to other carbon sources and they have many advantages in making CDs. It is inexpensive, easy to get, green, and harmless. In addition, the production of CDs from natural biomass can turn low-value organic waste into valuable and useful materials. Organic precursors play an important role in the entire energy system.

6.2.1 APPLICATIONS

As a group of newly emerged fluorescent nanomaterials, CDs show tremendous potential for a variety of applications including chemical sensing,[12] biosensing,[13] bioimaging,[14] drug delivery,[15] photocatalysis,[16] and electrocatalysis,[17] due to its excellent photoluminescence (PL) properties and ease of modifying their optical properties through doping and functionalization. Traditional fluorescent labeling materials have the disadvantages of a complex synthesis process and are easily self-assembled in the aqueous phase, which limits their application. Compared to conventional fluorescent dyes, CDs have higher stability and are more easily dispersed with water. Over the past few years, there has been tremendous progress in the use of renewable, inexpensive, and green resources not only addressing the urgent need for large-scale synthesis of CDs but also promoting the development of sustainable applications.

6.2.1.1 CHEMICAL SENSING

An interesting application of CDs is in the field of sensing. In principle, any PL changes including PL intensity wavelength, anisotropy, or a lifetime in the presence of different concentrations of specific analytes are likely for application in fluorescence-based sensing. It is done with the interaction of analytes and chelating agents on the surface of CDs designed through the introduction of functional groups (either from the precursors or solvent) during the synthesis process, postfunctionalization or integration with other molecules such as quenchers or fluorophores.[18] CD-based fluorescence sensing can be attributed to various mechanisms including photo-induced electron transfer,[19] fluorescence resonance energy transfer,[20] inner filter effect,[21] electron transfer,[22] aggregation-induced emission enhancement effect,[23] aggregation-induced emission quenching effect,[24] static quenching effect,[25] and dynamic quenching effect.[26] These principles make CDs a great candidate for the detection of heavy metals, cations, anions, biomolecules, biomarkers, nitroaromatic explosives, pollutants, vitamins, and drugs.

Fe^{3+} are the most common detected ions since the functional groups such as amino groups, carboxyl groups, and hydroxyl groups are generally present on the surface of CDs and thus can be effectively combined with iron ions. Yang et al. reported a procedure for selective detection of Fe^{3+} from honey.[27] Coriander leaves,[28] egg white,[29] garlic,[30] sugar cane molasses,[31] rose-heart radish,[32] grape peel,[33] and other diverse biomass are used to prepare CDs for iron ion detection. The detection of heavy metals such as Hg^{2+} is crucial because of their hazardous effect on the environment and human health. Green CDs have been employed for selective and sensitive Hg^{2+} detection on various occasions. Various precursors such as pomelo peel,[34] flour,[35] strawberry juice,[36] tea,[26] urine,[37] hair,[38] edible mushrooms,[39] etc. are used to detect mercury ion from the water bodies. CDs prepared from biomass precursors such as bamboo leaves,[40] *Ocimum sanctum*,[41] pigskin,[42] peach gum,[43] and lemon peel[16] are widely used for the detection of other various cations, such as Cu^{2+}, Pb^{2+}, Co^{2+}, Au^{3+}, and Cr^{6+}.

Compared to cations detection, there are few reports about the use of biomass CDs for anion detection. Various cations can quench the fluorescence of biomass CDs. After adding some specific anions, the fluorescence was recovered because of the binding between cations and anions. This on–off phenomenon can be used to detect many different combinations of cations and anions including $(Cu^{2+}–S^{2-})$, $(Fe^{3+}–S_2O_3^{2-})$,[44] $(Fe^{3+}–PO_4^{3-})$, etc. and CDs detect some other anions by quenching mechanisms, for example, ClO^-,[45] CN^-, etc. Xu et al. synthesized blue-fluorescence CDs from potatoes for the detection of PO_4^{3-} ion.[46] In addition to detecting ions, CDs can also be used to detect molecules, including tartazine,[47] tetracycline,[48] glutathiones,[49] etc. Organic CDs have appreciable selectivity and sensitivity which may lead to the fabrication of devices for real-time detection.

6.2.1.2 BIOIMAGING

Bioimaging is an intriguing application for biomass CDs. The extremely small size of CDs means they can be taken up easily by cells which can be imaged for intracellular fluorescence. These biomass CDs have been used to culture living cells such as *HaCaT* cells,[50] *E. coli*,[51]

HepG2 cells[52] to test their potential for dual-model fluorescence/MR imaging. In addition to the effect of biocompatibility of biomass CDs, the preparation of CDs with various emission wavelengths also plays a vital role in bioimaging. Park et al. used mango fruit as a carbon source to synthesize CDs. The obtained CDs exhibited blue, green, and yellow fluorescence which were used as a probe for bioimaging.[53] Moreover, the size of biomass CDs also has a certain impact on cell imaging. The small size biomass CDs are more easily captured by cells and contribute to intracellular fluorescence imaging. Shi et al. used various plant petals as carbon precursors to prepare CDs for cellular imaging of human uterine cervical squamous cell carcinoma (A193).[54] Because of the specificity of cells, pH value is also a significant factor affecting biomass CDs bioimaging. Shuang et al. produced fluorescence CDs derived from leeks to study Cu^{2+} and pH sensing in cells. The CDs with pH-dependent behavior exhibited turn-on fluorescence as pH ranging from 3 to 11.[55] Although the development of CDs is still in its infancy, CDs have great potential in bioimaging and promote the development of biomedicine.

6.2.1.3   CATALYSIS

Catalysis is the next major area of application of organic CDs. Researchers always doped natural CDs with other materials and the as-composited CDs can be used in electrocatalysis and photocatalysis. Because of their photoluminescent properties and photoelectron transfer properties, CDs can be considered an active ingredient in the manufacture of high-performance photocatalysts. Heteroatom doping functionalization of CDs can adjust the bandgap and increase the quantum yield and have a certain catalytic performance of the CDs. Tyagi et al. prepared CDs from lemon peel waste using a facile and cost-effective process. The photocatalytic activity of the TiO_2-CDs composites was confirmed by immobilizing the synthesized CDs on the electrospun TiO_2 nanofibers and using methylene blue (MB) dye as a model pollutant. The photocatalytic activity of the TiO_2-CDs composite is approximately 2.5 times greater than that of the TiO_2 nanofibers.[16]

6.2.1.4   DRUG DELIVERY

Biomass CDs are highly fluorescent, excellent biocompatible, rapid cellular acquisition, and high stability, so they serve as a multifunctional vehicle for drug loading and release. D'souza et al. used dried shrimp as the source material of CDs and were reasonably made into a traceable drug delivery system for targeted delivery of crude in MCF-7 cells.[56] Mehta et al. synthesized CDs by using pasteurized milk as a carbon source. Then, they prepared lisinopril-loaded CDs by self-assembly of lisinopril on the surfaces of CDs.[57] The abundance of chemical groups such as amino or hydroxyl groups on CDs can promote their future functionalization. These highly biocompatible CDs can serve as a novel fluorescent tool for studying drug delivery.

6.2.1.5   SOLAR CELLS

The earth soaks up to 86 PW of solar energy every year and harnessing this energy using photovoltaic devices is one of the keys to making the world carbon neutral. Unfortunately, current photovoltaic devices are limited by their high manufacturing costs and low efficiency. One possible solution is to use organic CDs in a new generation of solar cells since they show excitation-dependent or excitation-independent fluorescence emission and size-dependent fluorescence emission. Zhang et al. reported a fluorescence quenching mechanism that significantly improved the conversion efficiency of CDs sensitized aqueous solar cells. CDs synthesized from grass were chosen as the test case to validate the principle proposed to be responsible for enhancing the fluorescence quenching.[58]

6.2.1.6   OTHER FIELDS

In addition to some of the above-mentioned significant and wide range of applications, there are some other applications such as light display materials, anti-counterfeiting materials, and confidential materials. Zhang et al. presented an investigation of CDs synthesized from tofu wastewater. This study is relevant to the application of fluorescent CDs as light display materials.[59] Liu et al. used hair as a carbon source to produce highly fluorescent CDs by a one-step pyrolysis treatment. These CDs are useful in fluorescent patterns, flat panel displays, and anti-counterfeiting labeling. The electron-donor capabilities of photoexcited CDs have clear potential in reduction reactions.[60]

Carbon dots have proven benefits over conventional QDs and organic fluorophores for various applications resulting from their easy synthesis, low cytotoxicity, and superior optical properties. However, biomass CDs with high quantum yields (QYs) are still less reported. In addition, improving sensitivity and selectivity is a challenge. Based on the unique characterization of CDs, various types of probes were fabricated to detect various metal ions, anions, small molecules, and macromolecules by observing the process of quenching or recovery of the fluorescence of CDs. Another exciting area for further development is specific targeting of cellular organelles with green CDs. The CDs-based bioimaging probes could be tuned for high intracellular photostability for long term imaging. Another important application is in the field of electronics. Organic CDs are the future of electronic devices with their high performances including biocompatibility, nonphotoblinking, and excellent fluorescence properties. They have potential for use in various applications including biosensors, organic light-emitting devices, energy storage devices, and organic photovoltaics. These qualities have inspired great vision in the preparation and application of biomass CDs, helping to achieve green chemistry. The remarkable progress in developing different organic CDs for a variety of applications is actually what separates them from other carbon structures and makes them the next generation material.

6.3 CARBON NANOTUBES

CNTs, due to their extraordinary properties, find applications in various fields. Owing to the superior performance of CNTs, there are hundreds of articles published from a multidisciplinary approach that involve physics, chemistry, biology, biochemistry, electronics, and materials science.[61] CNTs have excellent electronic,[62] optical,[63] thermal,[64] chemical, and mechanical characteristics,[65,66] which makes them a valuable material for environment friendly applications. In this era of environmental degradation, developing an alternative to nonrenewable fossil fuel-related energy economy is of vital importance. Hence, this section focuses on the applications of CNTs synthesized from natural precursors. There is a strong relationship between the morphology of CNT synthesized and its application. This leads to variation in properties and the application of CNTs synthesized from natural precursors. In short, there will be a special and unique purpose for each natural hydrocarbon derived CNTs.

The major applications dealt in this fraction of chapter involve wastewater treatment, application in solar cells, capacitors, and hydrogen storage. Prevention of environmental degradation, improvement of public health, energy efficiency optimization, remediation, wastewater reuse, and pollutant transformation are the other applications of CNTs (Fig. 6.2). But in the following section, the main focus is given to applications of green synthesized CNTs.

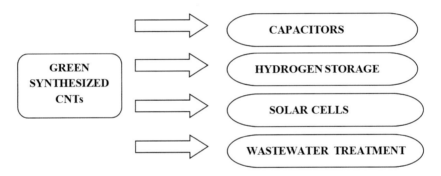

FIGURE 6.2   Schematic illustration of applications of green synthesized CNTs.

6.3.1   APPLICATIONS

6.3.1.1   WASTEWATER TREATMENT

Due to the high surface-to-volume ratio and a well-arranged distribution of pore size, CNTs have excellent sorption capacity. This ability of CNTs is a boon as it can be utilized for the removal of toxic heavy metals from water.[67–69] Water is a necessity for all living beings which is polluted to such an extent that the next war will be for water as predicted. Adsorption capacity of CNTs depends on the sorbate and the functional groups on the surface. Several heavy metals and other organic pollutants can be removed from wastewater by utilizing CNTs as adsorbents.[70–76] But a major limitation in using those chemically synthesized and modified CNT in the water treatment process is that those which are used to remove pollution can become pollution in the long run. This necessitates the need for green synthesized CNTs in the prevention of environmental degradation. Fortunately, there are successful reports on using sustainable CNTs for purification. Arsenic which is regarded as a toxic heavy metal ion was adsorbed on the surface of CNTs synthesized by using castor oil as a precursor.[77] This actually gives hope for the development of such CNTs in the future for sustainable development.

6.3.1.2 SOLAR CELLS

In order to meet the long term demand for energy and protect environmental balance, there is a high requirement for renewable energy sources. Application in photovoltaic devices was a major breakthrough contribution of CNTs. Generation of electricity in photovoltaic devices is through the conversion of photons absorbed from the sun. The commercially available silicon and semiconductor-based photovoltaic devices have several drawbacks including low stability under illumination and high cost.[78,79] Due to the affordability and noteworthy energy conversion CNTs are suitable alternatives in solar cell architectures. The mobility of the charge transfer increases due to the presence of delocalized π-electron system and the presence of nanoscale active surface area provides massive absorption of photons in CNTs. The fine alignment of CNTs enhances the photoconductivity on illumination.[80-82] The efficiency of solar cells can be enhanced by depositing CNTs on transparent conducting oxide. It has enormous advantages over other catalysts such as platinum because in those cases there are chances for degradation with time and possesses a greater threat. CNTs derived from olive oil proved to be a better substitute for such catalysts.[83]

6.3.1.3 CAPACITORS

Ultracapacitors or electrochemical double-layer capacitors which are energy storage devices working on the principle of double-layer effect shows enhanced activity when CNTs are used as electrodes.[84] CNTs with vertical alignment as electrodes in capacitors show superior power density, four times higher than traditional batteries, large lifespan, and greater energy density. All these enhancements are devoted to the remarkable conductivity and surface area of CNTs.[85–87] As a sustainable alternative, CNTs synthesized from natural hydrocarbons are noteworthy. The porous, curved, and hollow CNTs derived from sunflower seed hulls and sago have shown promising results as capacitors.[88] The cyclic voltammogram and related studies are the evidence for its excellent performance.

6.3.1.4 HYDROGEN ENERGY STORAGE

Hydrogen is considered as an efficient fuel due to its pollution-free nature, renewability, abundance, and energy efficiency.[89] CNTs and its modified forms have proved as a prominent future promising candidate for hydrogen storage in the past years.[90–94] Even though the accurate mechanism responsible for the adsorption of hydrogen by CNT is unknown still, it has been proposed that adsorption occurs mainly through two ways namely, physisorption and chemisorption. Physisorption in CNTs is through the trapping of hydrogen molecules inside the cylindrical structure of the nanotube whereas chemisorption occurs when hydrogen adsorbs exothermically to the top sites of carbon atoms on the wall of the tube. This promotes sp^3 like hybridization in the tube and results in expansion of diameter. Larger tube diameter leads to higher hydrogen storage capacity.[95] There are reports on the use of green synthesized CNT for the storage of hydrogen. Optimal hydrogen storage capacity achieved by CNT prepared using Kalonji seed oil as precursor opens a wide window of further discovery toward sustainable development.[96]

The main culprit of environmental degradation and pollution is considered as the over usage of fossil fuels. Green synthesis of carbon-based materials plays a revolutionizing role in the development of device applications to environment-friendly ones. CNTs have received significant research attention for their potential applications in different fields. Even though there are countless applications for CNTs since its discovery, the focus has been given to the applications of CNTs synthesized from green precursors. Considering today's scenario, stepping toward greener method is the only route to sustainable future development.

6.4 GRAPHENE/GRAPHENE OXIDE/REDUCED GRAPHENE OXIDE (rGO)

Nanotechnology is socially relevant in the field of biomedical applications and environmental remediation. Unusual electronic structure and large surface area of carbon-based nanomaterials are its various applications. For water purification, carbon is one of the most versatile materials. According to Vedic literature, charcoal is used for water purification. Activated carbon obtained from plant sources is also used as a water purifier. Graphene is one of the major allotrope of carbon and basic structure of all graphitic materials, graphene has a 2D arrangement of carbon atom with honeycomb-like lattice structure. Graphene shows several unusual properties, due to these properties it is applicable to DNA sensing, drug delivery, biomedicine, neurogeneration, tumor therapy, deep brain stimulation, and water purification by the effective adsorption of heavy metal ion, pesticides, and dyes from water.[97,100,105] Recently, graphene is research material for the photothermal treatment of Alzheimer's disease. The presence of some functional groups like hydroxyl, carbonyl, and epoxy group is the major difference between graphene and its derivates including GO and rGO in addition to the application of graphene, GO is used for anti-HIV, antibacterial activity, etc.[100] Graphene and GO are applicable for gene therapy, bioimaging, drug delivery, and as a scaffold for cell culture in tissue engineering, biosensors, and antibacterial composites (Fig. 6.3).[106]

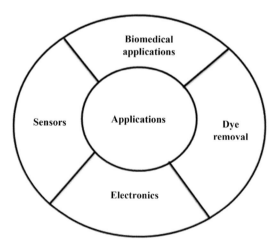

FIGURE 6.3   Schematic illustration of applications of graphene, GO, and rGO.

6.4.1 APPLICATIONS

Green synthesized graphene and its derivatives are applicable to various fields of science and technology. The quality and quantity of green-based products are equal or higher compared to the product obtained by the chemical route. One of the major applications of nanoparticles is drug delivery, the major advantage of this application is that the nanoparticles prepared for this purpose attracted only the diseased cells that will lead to the reduction of side effects and improve the efficacy (Fig. 6.3).

6.4.1.1 BIOMEDICAL APPLICATIONS

Gelatin–graphene nanosheet (GNS) is obtained by the education of GO using gelatin. Thus obtained Gelatin–GNS shows lots of applications in the biomedical field. For the purpose of drug delivery, gelatin–GNS should act as an excellent drug carrier. Mainly the anticancer drugs are delivered by this nanoparticle with maximum efficiency.[104] The drug-carrying capacity of graphene and its derivatives are due to its presence of 2D nanosheet. Heparin-rGO is used as an anticoagulant; it will inhibit the blood coagulation by both extrinsic and intrinsic pathways. But hydrazine-rGO inhibits coagulation only in the intrinsic pathway, and also in addition to the anticoagulation activity, heparin-rGO is capable of strong binding with anticancer drug. In biomedical field, heparin-rGO is a valuable material due to its anticoagulant activity and biocompatibility.[110] Bacillus marisflavi biomass is used as a reducing agent for the reduction of GO, bacterially reduced graphene oxide (B-rGO) is an efficient nanomaterials for application in effective nanotherapy of cancer cells. The as-prepared rGO are capable of various biomedical applications.[98] The rGO obtained by the effective reduction using glucose is used for photothermal LNCaP prostate cancer treatment in the presence of Fe catalyst. This photothermal therapy is more efficient than hydrazine-rGO. The synthesized glucose-rGO sheets material for photothermal cancer therapies.[109]

6.4.1.2 DYE REMOVAL

The rGO from aloe vera plant extract is used for the dye removal with maximum efficiency. About 98% dye removal is achieved by rGO, but in the case of GO only 80% of the dye is removed. rGO shows high adsorption toward MB because of the strong pi–pi interaction between MB and rGO surface. Effective surface area of rGO is the reason for the higher percentage of dye removal compared to GO.[99] The adsorption capacity of rGO is applicable for the wastewater treatment, mainly removal of organic pollutants. From the recyclability study, we can confirm that this green synthesized rGO shows excellent desorption capacity.[99] The rGO obtained by the reduction of GO using sugarcane bagasse, this B-rGO is capable of removing 92% MB when the rGO stabilized with ascorbic acid is removing 77% of methyl green dye.[108] The rGO obtained by the reduction of GO by the help of Vitis Vinifera (grapes) is used as a dye adsorbing material, specially malachite green solution, about 50% of dyes are removed within 5 min.[101]

6.4.1.3   SENSING

Hydrogen peroxide is a very essential component of many biological processes; H_2O_2 acts as an intermediate for lots of enzymatic reactions. So, to detect the presence of hydrogen peroxide as one of the important needs of the science world for this purpose construction of a chemically modified electrode is essential for sensing H_2O_2. Porous graphene nanosheet obtained from ficus-iacertiolia fruit acts as an excellent H_2O_2 sensor.[101] rGO obtained by the reduction of GO using rose water shows good electrochemical activity toward nicotinamide adenine dinucleotide, catechol, immobilized glucose oxidase. The glassy carbon electrode modified with rGO is used for this study.[107]

6.4.1.4   ELECTRONIC APPLICATIONS

The rGO derived from the reduction of GO using polyphenol extracts of Eucalyptus bark is used for the fabrication of high-performance supercapacitors. Cyclic voltammetry and galvanostatic charge–discharge studies show that the E-graphene supercapacitor has a high specific capacitance.[102] *Halomonas eurihalina* and *Halomonas maura* bacterias are used as reducing agents for the reduction of GO, the rGO obtained from *Halomonas eurihalina* is called as ERGO and those from the bacteria *Halomonas maura*is known as MRGO. Both of these materials are employed in green electronics.[112]

The green synthesized graphene, GO, and rGO are applicable in different fields like wastewater treatment, biomedical field, etc. In the future, graphene and its derivatives are some of the promising materials in the science world and its applications are useful for the society. The efficiency of green synthesized material is very high compared to the material obtained by chemical route. So, in the future development of green method with green precursors is one of the major challenges of the research field, but this method provides lots of advantages to our society. In the green method, there is no use of any toxic chemicals, which are harmful to human beings and nature, and also the green synthesized products are mainly applicable to pollutants removal and biomedical field.

6.5 FULLERENES

The third allotropic form of carbon after diamond and graphite, fullerene has won Nobel Prize of the year 1996 in its discovery by Harold W. Kroto, Robert F. Curl, and Richard E. Smalley. It consists of 20 hexagonal and 12 pentagonal rings as the foundation of an icosahedral symmetry closed cage structure. The structure of fullerene is only one of its kinds in many aspects including the fact that the molecules are edgeless, chargeless, and have no boundaries, no dangling bonds, and no unpaired electron. All these characteristics make fullerenes exceptional from other crystalline structures such as graphite or diamond which have edges with dangling bonds and electrical charges.[113] There are two bond lengths in C_{60} molecule, double bonds are considered as the 6:6 ring bonds which are smaller than 6:5 bonds. Poor electronic delocalization in fullerene is responsible for the behavior of fullerenes as electron-deficient alkenes.

Fullerenes have gained incredible importance in nanoscience due to their outstanding properties and valuable applications. Fullerenes have been produced from graphite,[114–117] burning of benzene, combustion of hydrocarbon,[118] and pyrolysis of naphthalene.[119] The high cost and harmful precursors in synthesis are major concerns limiting its application in various fields. Therefore, it is essential to search better alternatives for the production of fullerene using safer and greener sources. Unfortunately, there are only very few reports on green synthesized fullerenes till date.

Fullerenes were successfully synthesized from coal, coke, and camphor in the early 90s but the applications of those were not fully established.[120–122]

6.5.1 APPLICATIONS

The uniqueness of fullerene in physical and chemical properties led to a wide range of applications in almost all fields. Fullerene plays an excellent role in various areas such as organic photovoltaics, as antioxidants and biopharmaceuticals, as polymer additives, as catalysts, as water purification and biohazard protection catalysts, in portable power devices, as vehicle compounds, and for medical purposes. The size, hydrophobicity, electronic configuration, and 3D ability are some of the most attractive features of fullerene which have brought them to the front burner of medicinal chemistry.[123–126] All the interesting physical and chemical properties such as electrical properties, capacity to form endohedral, and exohedral derivatives, electron- accepting nature, mechanical strength, and negligible biotoxicity of fullerenes indirectly conveys the capability of fullerenes in several technological applications.

Fullerenes have the potential for various applications in day-to-day life. On hydrogenation, fullerenes form a C-H bond which has strength lower than that of the C-C bond. As a result, when they are heated, the C-H bond breaks preserving the typical structure of fullerene. This points out the scope of fullerenes as ideal molecules for hydrogen storage. Due to the reaction of fullerenes with free radicals that are harmful to the body, fullerenes hold grand promise in the pharmaceutical sector as antioxidants.[127] New copolymers of specific physical and mechanical properties can be created by adding fullerene to a polymer. When hydrophilic moieties are added to the hydrophobic C_{60} fullerene, it becomes soluble in water which can be utilized for transporting drugs to the cells. There are some reports in which the functionalized fullerenes are regulated to deliver drug gradually for getting maximum healing effect.[128]

Even though there are several reports for the applications of fullerenes in various fields, those which involve the use of green synthesized fullerenes are negligible. Most of the applications of fullerenes discussed are those which are prepared from fossil fuel and petroleum products. The two major restrictions of this approach are: there is a fear of depletion of these nonrenewable resources in the near future and it can affect the ecological balance of nature since it is not environment friendly. Therefore, it is high time for switching to green precursors for the development of fullerenes.

6.6   CARBON NANODIAMONDS (CNDs)

There are several applications attributed to CNDs including thermal, electrochemical, mechanical, cosmetic, and biomedical applications. But here we briefly described the applications of CNDs synthesized via green routes. The main highlight of the description accounts for the biomedical applications of CNDs. It briefly describes drug delivery, anticancer therapy, gene delivery, antimicrobial agents, bone and tissue implants, and bioimaging.

6.6.1   APPLICATIONS

6.6.1.1   DRUG DELIVERY

The well designed drug incorporation strategy aims targeted drug delivery and controlled release of drugs which improves therapeutic efficiency. The surface modification of nanodiamond results in the formation of covalent or electrostatic binding capacity to the bioactive molecules which results in prominent drug delivery. This application is strictly concentrating on anticancer therapy which comprises the direct or indirect influence of nanodiamonds for the target-specific delivery of the drug. The important aspect of the CND is the enhanced efficacy and nontoxic effect while act as a drug carrier. The pH-responsive of a drug delivery system with the therapeutic agent cisplatin (CP) was reported. One of the anticancer drug CP was carefully loaded on the nanodiamond based on adsorption and complexation property. The CP loaded on CND get released very effectively at a pH lower than 7.4 of the phosphate buffer solution.[129] The conjugation of doxorubicin (Dox) on small-sized CND, which sustain the potency of the drug.[130] Chow et al. studied the efficiency of nanodiamond conjugated therapeutic agent for the liver and mammary cancer in mouse models. The formation of nanodiamond and Dox complex makes an observable increase in cancer cell growth inhibition compared with free drugs. So it suggests the inevitable role of nanodiamond conjugated drug for the pronounced treatment of cancer cells.[131,132] Huang et al. reported the gathering of CND on CND-multilayer nanofilm with positively charged poly-L-lysine with dexamethasone as an anti-inflammatory transfer to the murine macrophages.[133] The most important applications of drug delivery and targeting are tabulated in Table 6.1.

TABLE 6.1   Applications of Nanodiamond in Drug Delivery and Targeting.

Therapeutic agents	Nature of CND	Applications	References
CP	ND-CP composite	pH-responsive release of CP and active in vitro against human cervical cancer cells.	[129]
Dexamethasone	CND (2–8 nm)-multilayer nanofilm with positively charged poly-L-lysine	Monitor the inflammation of RAW 264.7 murine macrophage using genetic analysis.	[133]
Dexamethasone and 4-hydroxy tamoxyfen	CND-drug complex	Enhance water dispersion of water insoluble drugs.	[134]
Doxorubicin HCl (Dox)	Nanodiamonds (2–8 nm)	Improve anticancer efficacy by reducing drug efflux-based chemoresistance	[130, 132, 135]
Doxorubicin HCl (Dox)	Microfilm architecture consists of Dox-CND (2–8 nm) conjugates	Localized, stable, and continues slow release of Dox for at least 1 month.	[136]
Folic acid (FA)	Fluorescent CND-FA conjugate	Receptor-mediated targeting of cancer cells.	[137]
Paclitaxel	CND (3–5 nm)-paclitaxel conjugate	In vitro mitotic arrest and apoptosis in A549 human lung carcinoma cells.	[138]

6.6.1.2   BIOIMAGING

Bioimaging is an advanced technique which promotes the in vivo imaging of biological process to molecular and cellular levels. It offers precise tracking of metabolites and is widely used as biomarkers for the identification of diseases. The biocompatible and nontoxicity of cells are the important criteria for offering CND as suitable for bioimaging applications. The single CND inside the Hela cell was tracked in a time gap of more than 200 s without any reduction in intensity.[137] The detection of even the single nitrogen-vacancy (NV) in living cells can be effectively done with the stimulated emission depletion (STED) spectroscopy. The detailed observation of the intracellular activities provided by the single CND technique increased the need of CND for improving the applications of CND in nanomedicine and nanobiology. CND also provides a better role in targeted cell imaging. The development of core-shell hybrid CND for targeted cell imaging is based on the property of aggregation of CND in biological solutions such as buffers, media, and blood. The adsorption of transferring,[139] growth hormone,[140] and chlorotoxin-like peptide[141] on the surface of CND forms polypeptide layer due to the enhanced stability of CND-complex. The small molecules including folic acid (FA),[137] cyclic peptide RGD,[142] and also biocompatible polymers are used as targeting molecules forming stable complexes with CND. Among the available multifunctional microscopic techniques, STED imaging has a prominent lateral resolution of less than 10 nm. The resolution capacity may reduce below the limit of diffraction of light due to the depletion effect caused by the high power STED depletion laser on neighboring fluorophore molecules. Thus, it causes the photobleaching of fluorophore and affects the observation time. But due to the accountable photostability, CND are used as an excellent source for STED imaging. Earlier analysis of Arreyo Camejo et al. showed the use of red fluorescent CND resolved the single NV with a resolution of about 10 nm.[143] The contrasting of CND with a size of about 55 nm in mammalian epithelial cells using differential interference contrast microscopy are reported by Smith B R et al. as shown in Figure 6.4.[144]

6.6.1.3   BIOSENSING

The different applications of CND in quantum sensing include magnetic protein detection, magnetic resonance imaging (MRI), near field coupling, and nanoscale thermometer. The magnetic protein ferritin affects the electron spin resonance of near field coupled NV centers. Thus, the use of CND as a magnetic sensor for the successful determination of ferritin concentrations.[145] The optical polarization of spins of electrons in NV centers and transferring polarization to ^{13}C nuclei leads to hyperpolarization. This offers enhanced sensitivity and high nuclear spin lifetime results in the application of MRI.[146] The specific arrangement of molecules and nanoparticles by DNA origami opened a new way in near field coupling investigation. The conjugation of neutravidin with CND and biotin-labeled oligonucleotide was investigated. The spin resonance property of NV and estimation of DNA origami in the nanolevel and distance-dependent coupling of CND highlighted to manage the NV base quantum devices in future applications of near field coupling and quantum sensing. To monitor local temperature change, the development of CND in the nanoscale thermometer was investigated. The pronounced sensitivity, sturdy reaction to the local environment, and high range detection enhanced the need of nanoscale thermometer.[147]

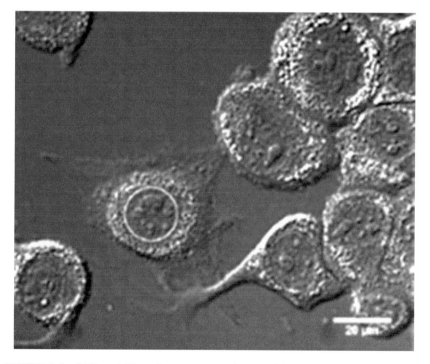

FIGURE 6.4 Differential interference contrast image of fixed 3T3 cells. CND particles are visible as bright rims around impermeable cell nuclei.
Source: Adapted with permission from Ref. [144].

The biomedical applications of CND synthesized using green routes are briefly described here. The biomedical uses of CND from the green method mainly constitute drug delivery, which is potentially applicable in medical challenging research such as cancer therapy, bioimaging, and biosensing.

6.7 CONCLUSIONS

The carbon nanomaterials attracted the attention of scientists from different threads of life which have pinched remarkable attention from fields ranging from chemistry, materials science, and engineering to condensed-matter physics and from both industry and academia. The unique properties of carbon nanomaterials such as graphene, CNT, carbon dots, etc. were inestimable in a variety of technologies like electronics, optics, energy storage, and many other applications. The previous 5 years have seen the massive advancement of carbon nanomaterials. They are widely utilized for different natural applications including fluorescence biosensing and bioimaging. In recent years, the focus on carbon nano-materials has been justified by a number of publications describing their unique properties that can be tailored to a particular type of application. Different determinants of the uniqueness of the carbon nanomaterials were recognized by Nobel prizes (2010) granted for the revelation of fullerene and graphene and the Kavli Prize granted for extraordinary commitments in propelling the information and comprehension of nanoscience and the disclosure of CNTs. These have without a doubt uncovered the uniqueness of these carbon-based materials which is unrivaled to some other ordinary noncarbon materials.

In spite of all the glaring possibilities of these carbon nanomaterials, a couple of disadvantages and constraints were likewise imagined in the mechanical scale production of some of these materials. In addition, the advanced application of carbon nanomaterials from organic precursors needs a higher degree of modification to achieve the goal. More advanced applications of these improved or functionalized carbon nanomaterials have to get on to the areas like nanomedicine for improved HIV drug ther-apies, DNA-based electronic devices, brain-inspired devices for artificial systems, super-powered bionic plants, light-seeking synthetic nano-robot, self-healable batteries, etc.

KEYWORDS

- **carbon-based nanomaterials**
- **carbon dots**
- **carbon nanotubes**
- **graphene**
- **fullerenes**
- **carbon nanodiamonds**
- **green applications**

REFERENCES

1. Aravind, A.; Mathew, B. An Electrochemical Sensor and Sorbent Based on Mutiwalled Carbon Nanotube Supported Ion Imprinting Technique for Ni(II) Ion from Electroplating and Steel Industries. *SN App. Sci.* **2018,** 1.
2. Sebastian, M.; Mathew, B. Carbon Nanotube-Based Ion Imprinted Polymer as Electrochemical Sensor and Sorbent for Zn(II) Ion from Paint Industry Wastewater. *Int. J. Polym. Anal. Ch.* **2017,** *23*, 18–28.
3. Sun, Y. P.; Zhou, B.; Lin, Y.; Wang, W.; Fernando, K. A. S.; Pathak, P.; Meziani, M. J.; Harruff, B. A.; Wang, X.; Wang, H.; Luo, P. G.; Yang, H.; Kose, M. E.; Chen, B.; Veca, L. M.; Xie, S. Y. Quantum-Sized Carbon Dots for Bright and Colorful Photoluminescence. *J. Am. Chem. Soc.* **2006,** *128*, 7756–7757.
4. Liu, R.; Wu, D.; Liu, S.; Koynov, K.; Knoll, W.; Li, Q. An Aqueous Route to Multicolor Photoluminescent Carbon Dots Using Silica Spheres as Carriers. *Angew Chem Int Ed.* **2009,** *48*, 4598–4601.
5. Lin, P.; Chen, J. W.; Chang, L. W.; Wu, J. P.; Redding, L.; Chang, H.; Yeh, T. K.; Yang, C. S.; Tsai, M. H.; Wang, H. J.; Kuo, Y. C.; Yang, R. S. H. Computational and Ultrastructural Toxicology of a Nanoparticle, Quantum Dot 705, in Mice. *Environ. Sci. Technol.* **2008,** *42*, 6264–6270.
6. Xu, X.; Ray, R.; Gu, Y.; Ploehn, H. J.; Gearheart, L.; Raker, K.; Scrivens, W. A. Electrophoretic Analysis and Purification of Fluorescent Single-Walled Carbon Nanotube Fragments. *J. Am. Chem. Soc.* **2004,** *126*, 12736–12737.
7. Miao, P.; Han, K.; Tang, Y.; Wang, B.; Lin, T.; Cheng, W. Recent Advances in Carbon Nanodots: Synthesis, Properties and Biomedical Applications. *Nanoscale* **2015,** *7*, 1586–1595.
8. Abbas, A.; Mariana, L. T.; Phan, A. N. Biomass-Waste Derived Graphene Quantum Dots and their Applications. *Carbon* **2018,** *140*, 77–99.
9. Wang, W.; Li, Y.; Cheng, L.; Cao, Z.; Liu, W. Water-Soluble and Phosphorus-Containing Carbon Dots with Strong Green Fluorescence for Cell Labelling. *J. Mater. Chem. B* **2014,** *2*, 46–48.

10. Chen, B.; Li, F.; Li, S.; Weng, W.; Guo, H.; Guo, T.; Zhang, X.; Chen, Y.; Huang, T.; Hong, X.; You, S.; Lin, Y.; Zeng, K.; Chen, S. Large Scale Synthesis of Photoluminescent Carbon Nanodots and Their Application for Bioimaging. *Nanoscale* **2013,** *5,* 1967.

11. Liu, H.; Ding, J.; Zhang, K.; Ding, L. Construction of Biomass Carbon Dots Based Fluorescence Sensors and Their Applications in Chemical and Biological Analysis. *Trends Anal. Chem.* **2019,** *118,* 315–337.

12. Guo, Y.; Wang, Z.; Shao, H.; Jiang, X. Hydrothermal Synthesis of Highly Fluorescent Carbon Nanoparticles from Sodium Citrate and Their Use for the Detection of Mercury Ions. *Carbon* **2013,** *52,* 583–589.

13. Bu, D.; Zhuang, H.; Yang, G.; Ping, X. An Immunosensor Designed for Polybrominated Biphenyl Detection Based on Fluorescence Resonance Energy Transfer (FRET) Between Carbon Dots and Gold Nanoparticles. *Sens. Actuators B* **2014,** *195,* 540–548.

14. Yu, C.; Li, X.; Zeng, F.; Zheng, F.; Wu, S. Carbon-Dot-Based Ratiometric Fluorescent Sensor for Detecting Hydrogen Sulfide in Aqueous Media and Inside Live Cells. *Chem. Commun.* **2013,** *49,* 403–405.

15. Mehta, V. N.; Chettiar, S. S.; Bhamore, J. R.; Kailasa, S. K.; Patel, R. M. Green Synthetic Approach for Synthesis of Fluorescent Carbon Dots for Lisinopril Drug Delivery System and Their Confirmations in the Cells. *J. Fluoresc.* **2016,** *27,* 111–124.

16. Tyagi, A.; Tripathi, K. M.; Singh, N.; Choudhary, S.; Gupta, R. K. Green Synthesis of Carbon Quantum Dots from Lemon Peel Waste: Applications in Sensing and Photocatalysis. *RSC Adv.* **2016,** *6,* 72423–72432.

17. Zhu, C.; Zhai, J.; Dong, S. Bifunctional Fluorescent Carbon Nanodots: Green Synthesis via Soy Milk and Application as Metal-Free Electrocatalysts for Oxygen Reduction. *Chem. Commun.* **2012,** *48,* 9367.

18. Edison, T. N. J. I.; Atchudan, R.; Shim, J. J.; Kalimuthu, S.; Ahn, B. C.; Lee, Y. R. Turn-Off Fluorescence Sensor for the Detection of Ferric Ion in Water Using Green Synthesized N-doped Carbon Dots and Its Bio-imaging. *J. Photochem. Photobiol.* **2016,** *158,* 235–242.

19. Miao, H.; Wang, L.; Zhuo, Y.; Zhou, Z.; Yang, X. Label-free Fluorimetric Detection of CEA Using Carbon Dots Derived from Tomato Juice. *Biosens. Bioelectron.* **2016,** *86,* 83–89.

20. Chatzimarkou, A.; Chatzimitakos, T. G.; Kasouni, A.; Sygellou, L.; Avgeropoulos, A.; Stalikas, C. D. Selective FRET-based Sensing of 4-nitrophenol and Cell Imaging Capitalizing on the Fluorescent Properties of Carbon Nanodots from Apple Seeds. *Sens. Actuator B-Chem.* **2018,** *258,* 1152–1160.

21. Purbia, R.; Paria, S. A Simple Turn on Fluorescent Sensor for the Selective Detection of Thiamine Using Coconut Water Derived Luminescent Carbon Dots. *Biosens. Bioelectron.* **2016,** *79,* 467–475.

22. Yuan, Y. H.; Liu, Z. H.; Li, R. S.; Zou, H. Y.; Lin, M.; Liu, H.; Huang, C. Z. Synthesis of Nitrogen-Doping Carbon Dots with Different Photoluminescence Properties by Controlling the Surface States. *Nanoscale* **2016,** *8,* 6770–6776.

23. Chen, B. B.; Li, R. S.; Liu, M. L.; Zhang, H. Z.; Huang, C. Z. Self-Exothermic Reaction Prompted Synthesis of Single-Layered Graphene Quantum Dots at Room Temperature. *Chem. Commun.* **2017,** *53,* 4958–4961.

24. Wang, N.; Liu, Z. X.; Li, R. S.; Zhang, H. Z.; Huang, C. Z.; Wang, J. The Aggregation Induced Emission Quenching of Graphene Quantum Dots for Visualizing the Dynamic Invasions of Cobalt (ii) into Living Cells. *J. Mater. Chem. B* **2017**, *5*, 6394–6399.

25. Yu, J.; Song, N.; Zhang, Y.; Zhong, S.; Wang, A.; Chen, J. Green Preparation of Carbon Dots by Jinhua Bergamot for Sensitive and Selective Fluorescent Fetection of Hg^{2+} and Fe^{3+}. *Sens. Actuators B* **2015**, *214*, 29–35.

26. Wei, J.; Liu, B.; Yin, P. Dual Functional Carbonaceous Nanodots Exist in a Cup of Tea. *RSC Adv.* **2014**, *4*, 63414–63419.

27. Yang, X.; Zhuo, Y.; Zhu, S.; Luo, Y.; Feng, Y.; Dou, Y. Novel and Green Synthesis of High-Fluorescent Carbon Dots Originated from Honey for Sensing and Imaging. *Biosens. Bioelectron.* **2014**, *60*, 292–298.

28. Sachdev, A.; Gopinath, P. Green Synthesis of Multifunctional Carbon Dots from Coriander Leaves and Their Potential Application as Antioxidants, Sensors and Bioimaging Agents. *Analyst* **2015**, *140*, 4260–4269.

29. Ye, Q.; Yan, F.; Luo, Y.; Wang, Y.; Zhou, X.; Chen, L. Formation of N, S-co-doped Fluorescent Carbon Dots from Biomass and Their Application for the Selective Detection of Mercury and Iron Ion. *Spectrochim. Acta Part A* **2017**, *173*, 854–862.

30. Sun, C.; Zhang, Y.; Wang, P.; Yang, Y.; Wang, Y.; et al. Synthesis of Nitrogen and Sulfur Co-doped Carbon Dots from Garlic for Selective Detection of Fe^{3+}. *Nanoscale Res. Lett.* **2016**, *11*, 110.

31. Huang, G.; Chen, X.; Wang, C.; Zheng, H.; Huang, Z.; Chen, D.; Xie, H. Photoluminescent Carbon Dots Derived from Sugarcane Molasses: Synthesis, Properties, and Applications. *RSC Adv.* **2017**, *7*, 47840–47847.

32. Liu, W.; Diao, H.; Chang, H.; Wang, H.; Li, T.; Wei, W. Green Synthesis of Carbon Dots from Rose-Heart Radish and Application for Fe^{3+} Detection and Cell Imaging. *Sens. Actuators B* **2017**, *241*, 190–198.

33. Xu, J., Lai, T., Feng, Z., Weng, X., Huang, C. Formation of Fluorescent Carbon Nanodots from Kitchen Wastes and their Application for Detection of Fe^{3+}. *Luminescence.* **2014**, *30*, 420–424.

34. Wang, Y.; Jiang, X. Synthesis of Cell-Penetrated Nitrogen-Doped Carbon Dots by Hydrothermal Treatment of Eggplant Sepals. *Sci. China: Chem.* **2016**, *59*, 836–842.

35. Qin, X.; Lu, W.; Asiri, A. M.; Al-Youbi, A. O.; Sun, X. Microwave-Assisted Rapid Green Synthesis of Photoluminescent Carbon Nanodots from Flour and their Applications for Sensitive and Selective Detection of Mercury (ii) Ions. *Sens. Actuators B* **2013**, *184*, 156–162.

36. Huang, H.; Lv, J. J.; Zhou, D. L.; Bao, N.; Xu, Y.; Wang, A. J.; Feng, J. J. One-Pot Green Synthesis of Nitrogen-Doped Carbon Nanoparticles as Fluorescent Probes for Mercury Ions. *RSC Adv.* **2013**, *3*, 21691–21696.

37. Essner, J. B.; Laber, C. H.; Ravula, S.; Polo-Parada, L.; Baker, G. A. Pee-Dots: Biocompatible Fluorescent Carbon Dots Derived from the Upcycling of Urine. *Green Chem.* **2016**, *18*, 243–250.

38. Guo, Y.; Zhang, L.; Cao, F.; Leng, Y. Thermal Treatment of Hair for the Synthesis of Sustainable Carbon Quantum Dots and the Applications for Sensing Hg^{2+}. *Sci Rep.* **2016**, 6.

39. Venkateswarlu, S.; Viswanath, B.; Reddy, A. S.; Yoon, M. Fungus-Derived Photoluminescent Carbon Nanodots for Ultrasensitive Detection of Hg^{2+} Ions and Photoinduced Bactericidal Activity. *Sens. Actuators B* **2018**, *258*, 172–183.

40. Liu, Y.; Zhao, Y.; Zhang, Y. One-Step Green Synthesized Fluorescent Carbon Nanodots from Bamboo Leaves for Copper(ii) Ion Detection. *Sens. Actuators B* **2014**, *196*, 647–652.

41. Kumar, A.; Chowdhuri, A. R.; Laha, D.; Mahto, T. K.; Karmakar, P.; Sahu, S. K. Green Synthesis of Carbon Dots from Ocimum sanctum for Effective Fluorescent Sensing of Pb^{2+} Ions and Live Cell Imaging. *Sens. Actuators B* **2017**, *242*, 679–686.

42. Wen, X.; Shi, L.; Wen, G.; Li, Y.; Dong, C.; Yan, J.; Shuang, S. Green and Facile Synthesis of Nitrogen-Doped Carbon Nanodots for Multicolor Cellular Imaging and Co^{2+} Sensing in Living Cells. *Sens. Actuators B* **2016**, *235*, 179–187.

43. Liao, J.; Cheng, Z.; Zhou, L. Nitrogen-Doping Enhanced Fluorescent Carbon Dots: Green Synthesis and Their Applications for Bioimaging and Label-Free Detection of Au^{3+} Ions, *ACS Sustain. Chem. Eng.* **2016**, *4*, 3053–3061.

44. Vandarkuzhali, S. A. A.; Jeyalakshmi, V.; Sivaraman, G.; Singaravadivel, S.; Krishnamurthy, K. R.; Viswanathan, B. Highly Fluorescent Carbon Dots from Pseudo-Stem of Banana Plant: Applications as Nanosensor and Bio-Imaging Agents, *Sens. Actuators B* **2017**, *252*, 894–900.

45. Yin, B.; Deng, J.; Peng, X.; Long, Q.; Zhao, J.; Lu, Q.; Chen, Q.; Li, H.; Tang, H.; Zhang, Y. Green Synthesis of Carbon Dots with Down- and Up-Conversion Fluorescent Properties for Sensitive Detection of Hypochlorite with a Dual-Readout Assay. *Analyst* **2013**, *138*, 6551–6557.

46. Xu, J.; Zhou, Y.; Cheng, G.; Dong, M.; Liu, S.; Huang, C. Carbon Dots as a Luminescence Sensor for Ultrasensitive Detection of Phosphate and Their Bioimaging Properties. *Luminescence* **2015**, *30*, 411.

47. Chatzimitakos, T.; Kasouni, A.; Sygellou, L.; Avgeropoulos, A.; Troganis, A.; Stalikas, C. Two of a Kind but Different: Luminescent Carbon Quantum Dots from Citrus Peels for Iron and Tartrazine Sensing and Cell Imaging. *Talanta* **2017**, *175*, 305–312.

48. Miao, H.; Wang, Y.; Yang, X. Carbon Dots Derived from Tobacco for Visually Distinguishing and Detecting Three Kinds of Tetracyclines, *Nanoscale* **2018**, *10*, 8139–8145.

49. Lu, X.; Liu, C.; Wang, Z.; Yang, J.; Xu, M.; Dong, J.; Wang, P.; Gu, J.; Cao, F. Nitrogen Doped Carbon Nanoparticles Derived from Silkworm Excrement as On-Off-On Fluorescent Sensors to Detect Fe(III) and Biothiols, *Nanomaterials* **2018**, *8*, 443–455.

50. Alam, A. M.; Park, B. Y.; Ghouri, Z. K.; Park, M.; Kim, H. Y. Synthesis of Carbon Quantum Dots from Cabbage with Down- and Up-Conversion Photoluminescence Properties: Excellent Imaging Agent for Biomedical Applications. *Green Chem.* **2015**, *17*, 3791–3797.

51. Das, P.; Bose, M.; Ganguly, S.; Mondal, S.; Das, A. K..; Banerjee, S.; Das, N. C. Green Approach to Photoluminescent Carbon Dots for Imaging of Gram-Negative Bacteria Escherichia coli. *Nanotechnology* **2017**, *28*, 195501.

52. Li, L.-S.; Jiao, X.-Y.; Zhang, Y.; Cheng, C.; Huang, K..; Xu, L.; Green Synthesis of Fluorescent Carbon Dots from Hongcaitai for Selective Detection of Hypochlorite and Mercuric Ions and Cell Imaging. *Sens. Actuators B* **2018**, *263*, 426–435.

53. Jeong, C. J.; Roy, A. K..; Kim, S. H.; Lee, J.-E.; Jeong, J. H.; In, I.; Park, S. Y. Fluorescent Carbon Nanoparticles Derived from Natural Materials of Mango Fruit for Bioimaging Probes. *Nanoscale* **2014**, *6*, 15196–15202.

54. Shi, L.; Li, Y.; Li, X.; Wen, X.; Zhang, G;, Yang, J.; Shuang, S. Facile and Eco-Friendly Synthesis of Green Fluorescent Carbon Nanodots for Applications in Bioimaging, Patterning and Staining. *Nanoscale* **2015**, *7*, 7394–7401.

55. Shi, L., Li, Y.; Li, X.; Zhao, B.; Wen, X.; Zhang, G.; Shuang, S. Controllable synthesis of Green and Blue Fluorescent Carbon Nanodots for pH and Cu^{2+} Sensing in Living Cells. *Biosens. Bioelectron.* **2016**, *77*, 598–602.

56. D'souza, S. L.; Deshmukh, B.; Bhamore, J. R.; Rawat, K. A.; Lenka, N.; Kailasa, S. K. Synthesis of Fluorescent Nitrogen-Doped Carbon Dots from Dried Shrimps for Cell Imaging and Boldine Drug Delivery System. *RSC Adv.* **2016**, *6*, 12169–12179.

57. Mehta, V. N.; Chettiar, S. S.; Bhamore, J. R.; Kailasa, S. K.; Patel, R. M. Green Synthetic Approach for Synthesis of Fluorescent Carbon Dots for Lisinopril Drug Delivery System and Their Confirmations in the Cells. *J. Fluore.* **2016**, *27*, 111–124.

58. Zhang, H.; Wang, Y.; Liu, P.; Li, Y.; Yang, H. G.; An, T.; Zhao, H. A Fluorescent Quenching Performance Enhancing Principle for Carbon Nanodot-Sensitized Aqueous Solar Cells. *Nano Energy* **2015**, *13*, 124–130.

59. Zhang, J.; Wang, H.; Xiao, Y.; Tang, J.; Liang, C.; Li, F.; Xu, W. A Simple Approach for Synthesizing of Fluorescent Carbon Quantum Dots from Tofu Wastewater. *Nanoscale Res. Lett.* **2017**, 12.

60. Liu, S. S.; Wang, C. F.; Li, C.-X.; Wang, J.; Mao, L.-H.; Chen, S. Hair-Derived Carbon Dots Toward Versatile Multidimensional Fluorescent Materials. *J. Mater. Chem. C* **2014**, *2*, 6477–6483.

61. Siqueira, R.; Oliveira, O. N.; Carbon-Based Nanomaterials. *Nanostructures.* **2017**, 233–249.

62. Zhu, J.; Holmen, A.; Chen, D. Carbon Nanomaterials in Catalysis: Proton Affinity, Chemical and Electronic Properties, and their Catalytic Consequences. *Chem. Cat Chem.* **2013**, *5*, 378–401.

63. Karami, M.; Bahabadi, M. A.; Delfani, S.; Ghozatloo, A. A New Application of Carbon Nanotubes Nanofluid as Working Fluid of Low-Temperature Direct Absorption Solar Collector. *Sol. Energy Mater. Sol. Cells* **2014**, *121*, 114–118.

64. Esfe, M. H.; Saedodin, S.; Yan, W.-M.; Afrand, M.; Sina, N. Study on Thermal Conductivity of Water-Based Nanofluids with Hybrid Suspensions of CNTs/Al_2O_3 Nanoparticles. *J. Therm. Anal. Calorim.* **2016**, *124*, 455–460.

65. Mauter, M. S.; Elimelech, M. Environmental Applications of Carbon-Based Nanomaterials. *Environ. Sci. Technol.* **2008**, *42*, 5843–5859.

66. Thirumal, V.; Pandurangan, A.; Jayavel, R.; Krishnamoorthi, S.; Ilangovan, R. Synthesis of Nitrogen Doped Coiled Double Walled Carbon Nanotubes by Chemical Vapor Deposition Method for Supercapacitor Applications. *Curr. Appl. Phys.* **2016**, *16*, 816–825.

67. Pokhrel, L. R.; Ettore, N.; Jacobs, Z. L.; Zarr, A.; Weir, M. H.; Scheuerman, P. R.; Kanel, S. R.; Dubey, B. Novel Carbon Nanotube (CNT)-Based Ultrasensitive Sensors

for Trace Mercury(II) Detection in Water: A Review. *Sci. Total Environ.* **2017,** *574,* 1379–1388.

68. Anitha, K.; Namsani, S.; Singh, J. K. Removal of Heavy Metal Ions Using a Functionalized Single-Walled Carbon Nanotube: A Molecular Dynamics Study. *J. Phys. Chem. A.* **2015,** *119,* 8349–8358.

69. Tofighy, M. A.; Mohammadi, T. Nickel Ions Removal from Water by Two Different Morphologies of Induced CNTs in Mullite Pore Channels as Adsorptive Membrane. *Ceram. Int.* **2015,** *41,* 5464–5472.

70. Ma, J.; Yu, F.; Zhou, L.; Jin, L.; Yang, M.; Luan, J.; Tang, Y.; Fan, H.; Yuan, Z.; Chen, J. Enhanced Adsorptive Removal of Methyl Orange and Methylene Blue from Aqueous Solution by Alkali-Activated Multiwalled Carbon Nanotubes. *ACS Appl. Mater. Interfaces* **2012,** *4,* 5749–5760.

71. Wang, S.; Ng, C. W.; Wang, W.; Li, Q.; Hao, Z. Synergistic and Competitive Adsorption of Organic Dyes on Multiwalled Carbon Nanotubes. *Chem. Eng. J.* **2012,** *197,* 34–40.

72. Lou, J. C.; Jung, M. J.; Yang, H. W.; Han, J. Y.; Huang, W. H. Removal of Dissolved Organic Matter (DOM) from Raw Water by Single-Walled Carbon Nanotubes (SWCNTs). *J. Environ. Sci. Health Part A.* **2011,** *46,* 1357–1365.

73. Zhu, H.; Jiang, R.; Xiao, L.; Zeng, G. Preparation, Characterization, Adsorption Kinetics and Thermodynamics of Novel Magnetic Chitosan Enwrapping Nanosized -Fe_2O_3 and Multi-Walled Carbon Nanotubes with Enhanced Adsorption Properties for Methyl Orange. *Bioresour. Technol.* **2010,** *101,* 5063–5069.

74. Yang, W.; Lu, Y.; Zheng, F.; Xue, X.; Li, N.; Liu, D. Adsorption Behavior and Mechanisms of Norfloxacin onto Porous Resins and Carbon Nanotube. *Chem. Eng. J.* **2012,** *179,* 112–118.

75. Ncibi, M. C.; Sillanpää, M. Optimized Removal of Antibiotic Drugs from Aqueous Solutions Using Single, Double and Multi-Walled Carbon Nanotubes. *J. Hazard. Mater.* **2015,** *298,* 102–110.

76. Yang, Q.; Chen, G.; Zhang, J.; Li, H. Adsorption of Sulfamethazine by Multi-Walled Carbon Nanotubes: Effects of Aqueous Solution Chemistry. *Rsc Adv.* **2015,** *5,* 25541–25549.

77. Tripathi, S.; Sharon, M.; Maldar, N. N.; Shukla, J.; Sharon, M. Carbon Nanospheres and Nanotubes Synthesized from Castor Oil as Precursor; for Removal of As Dissolved in Water. *Arch. Appl. Sci. Res.* **2012,** *4,* 1788–1795.

78. Aberle, A. G. Surface Passivation of Crystalline Silicon Solar Cells: A Review. *Prog photovoltaics* **2000,** *8,* 473–487.

79. Green, M. A. Third Generation Photovoltaics: Solar Cells for 2020 and Beyond. *Physica E.* **2002,** *14,* 65–70.

80. Kamat, P. V. Meeting the Clean Energy Demand: Nanostructure Architectures for Solar Energy Conversion. *J. Phys. Chem. C* **2007,** *111,* 2834–2860.

81. Scarselli, M.; Scilletta, C.; Tombolini, F.; Castrucci, P.; De Crescenzi, M.; Diociaiuti, M.; Casciardi, S.; Gatto, E.; Venanzi, M.; Photon Harvesting with Multi Wall Carbon Nanotubes. *Superlattices Microstruct.* **2009,** *46,* 340–346.

82. Li, X.; Zhou, H.; Yu, P.; Su, L.; Ohsaka, T.; Mao, L. A. Miniature Glucose/O_2 Biofuel Cell with Single-Walled Carbon Nanotubes-Modified Carbon Fiber Microelectrodes as the Substrate. *Electrochem. Commun.* 2008, *10,* 851–854.

83. Sharma, S.; Ranjan, P.; Das, S.; Gupta, S.; Bhati, R.; Majumdar, A. Synthesis of Carbon Nanotube Using Olive Oil and its Application in Dye Sensitized Solar Cell. *Int. J. Renew. Energy Res.* **2012**, *2*, 274–279.

84. Rahman, G.; Najaf, Z.; Mehmood, A.; Bilal, S.; Shah, A. H. A.; Mian, S. A.; Ali, G. An Overview of the Recent Progress in the Synthesis and Applications of Carbon Nanotubes. *Carbon* **2019**, *5*, 3.

85. Sharma, P.; Ahuja, P. Recent Advances in Carbon Nanotube-Based Electronics. *Mater. Res. Bull.* **2008**, *43*, 2517–2526.

86. Thirumal, V.; Pandurangan, A.; Jayavel, R.; Krishnamoorthi, S.; Ilangovan, R. Synthesis of Nitrogen Doped Coiled Double Walled Carbon Nanotubes by Chemical Vapor Deposition Method for Supercapacitor Applications. *Curr. Appl. Phys.* **2016**, *16*, 816–825.

87. Signorelli, R.; Ku, D. C.; Kassakian, J. G.; Schindall, J. E. Electrochemical Double-Layer Capacitors Using Carbon Nanotube Electrode Structures. *Proc. IEEE* **2009**, *97*, 1837–1847.

88. Zhang, H. Y.; Niu, H. J.; Wang, Y. M.; Wang, C.; Bai, X. D.; Wang, S.; Wang, W. A Simple Method to Prepare Carbon Nanotubes from Sunflower Seed Hulls and Sago and Their Application in Supercapacitor. *Pigm. Resin Technol.* **2015**, *44*, 7–12.

89. Lim, K. L.; Kazemian, H.; Yaakob, Z.; Daud, W. R. W. Solid-State Materials and Methods for Hydrogen Storage: A Critical Review. *Chem. Eng. Technol.* **2010**, *33*, 213–226.

90. Durgun, E.; Ciraci, S.; Yildirim, T.; Functionalization of Carbon-Based Nanostructures with Light Transition-Metal Atoms for Hydrogen Storage. *Phys. Rev. B.* **2008**, *77*, 085405.

91. Lochan, R. C.; Head-Gordon, M. Computational Studies of Molecular Hydrogen Binding Affinities: The Role of Dispersion Forces, Electrostatics, and Orbital Interactions. *Phys. Chem. Chem. Phys.* **2006**, *8*, 1357–1370.

92. Surya, V. J.; Iyakutti, K.; Venkataramanan, N.; Mizuseki, H.; Kawazoe, Y. The Role of Li and Ni Metals in the Adsorbate Complex and Their Effect on the Hydrogen Storage Capacity of Single Walled Carbon Nanotubes Coated with Metal Hydrides, LiH and NiH_2. *Int. J. Hydrog. Energy* **2010**, *35*, 2368–2376.

93. Iyakutti, K.; Kawazoe, Y.; Rajarajeswari, M.; Surya, V. J. Aluminum Hydride Coated Single-Walled Carbon Nanotube as a Hydrogen Storage Medium. *Int. J. Hydrog. Energy.* **2009**, *34*, 370–375.

94. Surya, V. J.; Iyakutti, K.; Rajarajeswari, M.; Kawazoe, Y. Functionalization of Single-Walled Carbon Nanotube with Borane for Hydrogen Storage. *Physica E.* **2009**, *41*, 1340–1346.

95. Volpe, M.; Cleri, F. Chemisorption of Atomic Hydrogen in Graphite and Carbon Nanotubes. *Surf. Sci.* **2003**, *544*, 24–34.

96. Sharon, M.; Soga, T.; Afre, R.; Sathiyamoorthy, D.; Dasgupta, K.; Bhardwaj, S.; Sharon, M.; Jaybhaye, S. Hydrogen Storage by Carbon Materials Synthesized from Oil Seeds and Fibrous Plant Materials. *Int. J. Hydrog. Energy.* **2007**, *32*, 4238–4249.

97. Tavakoli, F.; Salavati-Niasari, M.; Badiei, A., and Mohandes, F. Green Synthesis and Characterization of Graphene Nanosheets. *Mater. Res. Bull.* **2015**, *63*, 51–57.

98. Gurunathan, S.; Woong Han, J.; Eppakayala, V. Kim, J. Green Synthesis of Graphene and its Cytotoxic Effects in Human Breast Cancer Cells. *Int. J. Nanomed.* **2013**, 1015.

99. Bhattacharya, G.; Sas, S.; Wadhwa, S.; Mathur, A.; McLaughlin, J.; Roy, S. S. Aloe Vera Assisted Facile Green Synthesis of Reduced Graphene Oxide for Electrochemical and Dye Removal Applications. *RSC Adv.* **2017,** *7,* 26680–26688.

100. Cherian, R. S.; Sandeman, S.; Ray, S.; Savina, I. N.; Ashtami, J.; Mohanan, P. V. Green Synthesis of Pluronic Stabilized Reduced Graphene Oxide: Chemical and Biological Characterization. *Colloids Surf. B Biointerfaces* **2019,** *179,* 94–106.

101. Liang, T.; Guo, X.; Wang, J.; Wei, Y.; Zhang, D.; Kong, S. Green Synthesis of Porous Graphene-like Nanosheets for High-Sensitivity Nonenzymatic Hydrogen Peroxide Biosensor. *Mater. Lett.* **2019,** *254,* 28–32.

102. Manchala, S.; Tandava, V. S. R. K.; Jampaiah, D.; Bhargava, S. K.; Shanker, V. Novel and Highly Efficient Strategy for the Green Synthesis of Soluble Graphene by Aqueous Polyphenol Extracts of Eucalyptus Bark and its Applications in High-Performance Supercapacitors. *ACS Sustainable Chem. Eng.* **2019,** *7,* 11612–11620.

103. Rowley-Neale, S. J.; Randviir, E. P.; Abo Dena, A. S.; Banks, C. E.; An Overview of Recent Applications of Reduced Graphene Oxide as a Basis of Electroanalytical Sensing Platforms. *Appl. Mater. Today* **2018,** *10,* 218–226

104. Liu, K.; Zhang, J. J.; Cheng, F. F.; Zheng, T. T.; Wang, C.; Zhu, J. J. Green and Facile Synthesis of Highly Biocompatible Graphene Nanosheets and its Application for Cellular Imaging and Drug Delivery. *J. Mater. Chem.* **2011,** *21,* 12034.

105. Gupta, S. S.; Sreeprasad, T. S.; Maliyekkal, S. M.; Das, S. K.; Pradeep, T. Graphene from Sugar and its Application in Water Purification. *ACS Appl. Mater. Interfaces* **2012,** *4,* 4156–4163.

106. Syama, S.; Mohanan, P. V. Comprehensive Application of Graphene: Emphasis on Biomedical Concerns. *Nano-Micro Lett.* **2019,** *11,* 1–31.

107. Haghighi, B.; Tabrizi, M. A. Green-Synthesis of Reduced Graphene Oxide Nanosheets using Rose Water and a Survey on their Characteristics and Applications. *RSC Adv.* **2013,** *3,* 13365.

108. Gan, L.; Li, B.; Chen, Y.; Yu, B.; Chen, Z. Green Synthesis of Reduced Graphene Oxide Using Bagasse and its Application in Dye Removal: A Waste-to-Resource Supply Chain. *Chemosphere* **2019,** *219,* 148–154.

109. Akhavan, O. E.; Aghayee, S.; Fereydooni, Y.; Talebi, A. The Use of a Glucose-Reduced Graphene Oxide Suspension for Photothermal Cancer Therapy. *J. Mater. Chem.* **2012,** *22,* 13773.

110. Wang, Y.; Zhang, P.; Fang Liu, C.; Zhan, L.; Fang Li, Y.; Huang, C. Z. Green and Easy Synthesis of Biocompatible Graphene for Use as an Anticoagulant. *RSC Adv.* **2012,** *2,* 2322.

111. Upadhyay, R. K.; Soin, N.; Bhattacharya, G.; Saha, S.; Barman, A.; Roy, S. S.; Grape Extract Assisted Green Synthesis of Reduced Graphene Oxide for Water Treatment Application. *Mater. Lett.* **2015,** *160,* 355–358.

112. Raveendran, S.; Chauhan, N.; Nakajima, Y.; Toshiaki, H.; Kurosu, S.; Tanizawa, Y.; Tero, R.; Y.; Yoshida, Hanajiri, T.; Maekawa, T.; Ajayan, P. M.; Sandhu, A.; Kumar, D. S. Ecofriendly Route for the Synthesis of Highly Conductive Graphene Using Extremophiles for Green Electronics and Bioscience. *Par. Par. Syst. Charact.* **2013,** *30,* 573–578.

113. Smalley, R. E. Self-Assembly of the Fullerenes. *Acc. Chem. Res.* **1992,** *25,* 98–105.

114. Kroto, H. W.; Heath, J. R.; O'Brien, S. C.; Curl, R. F.; Smalley, R. E. C_{60}: Buckminsterfullerene. *Nature* **1985**, *318*, 162–163.

115. Krätschmer, W.; Lamb, L. D.; Fostiropoulos, K.; Huffman, D. R. Solid C60: A New Form of Carbon. *Nature* **1990**, *347*, 354–358.

116. Taylor, R.; Hare, J. P.; Abdul-Sada, A. K.; Kroto, H. W. Isolation, Separation and Characterisation of the Fullerenes C_{60} and C_{70}: The Third Form of Carbon. *J. Chem. Soc. Chem. Commun.* **1990**, *20*, 1423.

117. Curl, R. F.; Smalley, R. E. Probing C_{60}. *Science* **1988**, *242*, 1017–1022.

118. Howard, J. B.; McKinnon, J. T.; Makarovsky, Y.; Lafleur, A. L.; Johnson, M. E. Fullerenes C_{60} and C_{70} in Flames. *Nature* **1991**, *352*, 139–141.

119. Taylor, R.; Langley, G. J.; Kroto, H. W.; Walton, D. R. M. Formation of C_{60} by Pyrolysis of Naphthalene. *Nature* **1993**, *366*, 728–731.

120. Mukhopadhyay, K.; Krishna, K. M.; Sharon, M. Fullerenes from Camphor: A Natural Source. *Phys. Rev. Lett.* **1994**, *72*, 3182–3185.

121. Pang, L. S. K.; Vassallo, A. M.; Wilson, M. A. Fullerenes from Coal: A Self-Consistent Preparation and Purification Process. *Energy Fuels* **1992**, *6*, 176–179.

122. Pang, L. S. K.; Prochazka, L. R. A.; Wilson, A.; Pallasser, R.; Fisher, K. J.; Gerald, J. D. F.; Taylor, G. H.; Willett, G. D.; Dance, I. G. Competitive Reactions During Plasma Arcing of Carbonaceous Materials. *Energy Fuels* **1995**, *9*, 38–44.

123. Kang, S.; Mauter, M. S.; Elimelech, M. C. Microbial Cycotoxicity of Carbon-Based Nanomaterials: Implications for the River Water and Waste Water Effluent. *Environ. Sci. Technol.* **2009**, *43*, 2684–2653.

124. Das, R.; Vecitis, C. D.; Schulze, A.; Cao, B.; Ismail, A. F.; Lu, X.; Ramakrishna, S. Recent Advances in Nanomaterials for Water Protection and Monitoring. *Chem. Soc. Rev.* **2017**, *46*, 6946–7020.

125. Burakov, A. E.; Galunin, E. V.; Burakov, I. V.; Kucherova, A. E.; Agarwal, S.; Tkachev, A. E.; Gupta, V. K. Adsorption of Heavy Metals on Conventional and Nanostructured Materials for Water Treatment Purposes: A Review. *Ecotoxicol. Environ. Saf.* **2018**, *148*, 702–712.

126. Barkry, R.; Vallant, R. M.; Najam-Ul, H.; Rawer, M.; Szabo, Z.; Huck, C. Y.; Bonn, G. K. Medicinal Applications of Fullerenes. *Inter. J. Nanomed.* **2007**, *2*, 639–649.

127. Yadav B. C.; Kumar, R. Structure, Properties and Applications of Fullerenes. *Int. J. Nanotechnol. Appl.* **2008**, *2*, 15–24.

128. Ranja, B.; Rainer M. V.; Muhammad N. H.; Matthias R.; Zoltan S.; Christian W. H.; Günther K. B. Medicinal Applications of Fullerenes. *Int. J. Nanomed.* **2007**, *2*, 639–649.

129. Guan, B.; Zou, F.; Zhi, J. Nanodiamond as the pH-Responsive Vehicle for an Anticancer Drug. *Small* **2010**, *14*, 1514–1519.

130. Huang, H.; Pierstorff, E.; Osawa, E.; Ho, D. Active Nanodiamond Hydrogels for Chemotherapeutic Delivery. *Nano Lett.* **2007**, *11*, 3305–3314.

131. Chow, E. K.; Zhang X. Q.; Chen, M.; Lam, R.; Robinson, E.; Huang, H.; Schaffer, D.; Osawa, E.; Goga, A.; Ho, D. Nanodiamond Therapeutic Delivery Agents Mediate Enhanced Chemoresistant Tumor Treatment. *Sci. Transl. Med.* **2011**, *73*, 21.

132. Merkel, T.; DeSimone, J. M. Dodging Drug-Resistant Cancer with Diamonds. *Sci. Transl. Med.* **2011**, *73*, 8.

133. Huang, E. H.; Pierstorff, E.; Osawa, E and Ho, D. Protein-Mediated Assembly of Nanodiamond Hydrogels into a Biocompatible and Biofunctional Multilayer Nanofilm, *ACS Nano* **2008**, *2*, 203–212.

134. Chen, M.; Pierstorff, E. D.; Lam, R.; Li, S. Y.; Huang, H.; Osawa, E and Ho, D. Nanodiamond-Mediated Delivery of Water-Insoluble Therapeutics. *ACS Nano* **2009**, *7*, 2016–2022.

135. Ma, X.; Zhao, Y.; Liang X. Nanodiamond Delivery Circumvents Tumor Resistance to Doxorubicin. *Acta Pharmacol. Sin.* **2011**, *5*, 543–544.

136. Lam R.; Chen, M.; Pierstorff, E.; Huang, H.; Osawa, E; Ho D. Nanodiamond-Embedded Microfilm Devices for Localized Chemotherapeutic Elution. *ACS Nano* **2008**, *10*, 2095–2102.

137. Zhang B.; Li, Y.; Fang, C. Y.; Chang, C. C.; Chen, C. S.; Chen, Y. Y.; Chang, H. C. Receptor-Mediated Cellular Uptake of Folate-Conjugated Fluorescent Nanodiamonds: A Combined Ensemble and Single-Particle Study. *Small* **2009**, *23*, 2716–2721.

138. Liu, K. K.; Zheng, W. W.; Wang, C.-C.; Chiu Y.-C.; Cheng, C.-L.; Lo, Y.-S.; Chen, C.; Chao, J. I. Covalent Linkage of Nanodiamond-Paclitaxel for Drug Delivery and Cancer Therapy. *Nanotechnology* **2010**, *31*, 315106.

139. Weng, M. F.; Chiang, S. Y.; Wang, N. S.; Niu, H. Fluorescent Nanodiamonds for Specifically Targeted Bioimaging: Application to the Interaction of Transferrin with Transferrin Receptor. *Diam. Relat. Mater.* **2009**, *18*, 587–591.

140. Cheng, C. Y.; Perevedentseva, E.; Tu, J. S.; Chung, P.-H.; Cheng, C. L.; Liu, K. K.; Chao, J. I.; Chen, P. H; Chang, C. C. Direct and In Vitro Observation of Growth Hormone Receptor Molecules in A549 Human Lung Epithelial Cells by Nanodiamond Labeling. *Appl. Phys. Lett.* **2007**, *16*, 163903.

141. Fu, Y.; An, N.; Zheng, S.; Liang, A and Li, Y. BmK CT-Conjugated Fluorescence Nanodiamond as Potential Glioma-Targeted Imaging and Drug. *Diam. Relat. Mater.* **2012**, *21*, 73–76.

142. Slegerova, J.; Hajek, M.; Rehor, I.; Sedlak, F.; Stursa, J.; Hruby, M.; Cigler, P. Designing the Nanobiointerface of Fluorescent Nanodiamonds: Highly Selective Targeting of Glioma Cancer Cells. *Nanoscale* **2015**, 415–420.

143. Lim, J. K.; Majetich, S. A.; Tilton, R. D. Stabilization of Super Paramagnetic Iron Oxide Core–Gold Shell Nanoparticles in High Ionic Strength Media. *Langmuir* **2009**, *23*, 13384–13393.

144. Smith, B. R.; Niebert, M.; Plakhotnik, T.; Zvyagin, A. V. Transfection and Imaging of Diamond Nanocrystals as Scattering Optical Labels. *J. Lumin.* **2007**, *127*, 260–263.

145. Ermakova, A.; Pramanik, G.; Cai, J. M.; Algara-Siller, G.; Kaiser, U.; Weil, T.; Tzeng, Y. K.; Chang, H. C.; McGuinness, L. P.; Plenio, M. B.; Naydenov, B.; Jelezko, F. Detection of a Few Metallo-Protein Molecules Using Color Centers in Nanodiamonds, *Nano Lett.* **2013**, *7*, 3305–3309.

146. Chen, Q.; Schwarz, I.; Jelezko, F.; Retzker, A.; Plenio, M. B. Optical Hyperpolarization of C^{13} Nuclear Spins in Nanodiamond Ensembles. *Phys. Rev. B* **2015**, *92*, 18.

147. Toyli, D. M.; Christle, D. J.; Alkauskas, A.; Buckley, B. B.; Van de Walle, C. G.; Awschalom, D. D. Measurement and Control of Single Nitrogen-Vacancy Center Spins Above 600 K. *Phys. Rev. B* **2012**, 3.

CHAPTER 7

Development of Green Technology Through Renewable and Sustainable Materials

REMYA VIJAYAN[1*], SIJO FRANCIS[2], and BEENA MATHEW[3]

[1]*School of Chemical Sciences, Mahatma Gandhi University, Kottayam, India*

[2]*Department of Chemistry, St. Joseph's College, Moolamattom, India*

[3]*School of Chemical Sciences, Mahatma Gandhi University, Kottayam, India*

Corresponding author. E-mail: remyavijayan88@gmail.com

ABSTRACT

In recent times, environmental and climate problems are one of the daunting issues across the world. Economic development without considering the environmental concerns causes the depletion of natural resources particularly water resources and air. This results in extensive ecological, financial, and social impairment on a global level. Therefore, it is very essential to enhance the usage of renewable resources and reduce the usage of nonrenewable resources to protect the environment for future generations. This reiterates the need to develop green technologies, which are required to achieve economic development without harming the environment. In this chapter, we discussed the synthesis of renewable and sustainable materials by different methods and their application in the development of green technology.

7.1   INTRODUCTION

Green technology aimed at the development and application of products, equipment, and methods to preserve the natural resources and environment, besides reducing the harmful impacts on the environment induced by human activities. The green technology methods must be sustainable, which means it will balance the fulfillment of human societal requirements without further depletion of the remaining environment and natural resources with the intention that these requirements can be met not only in the present time but in the indefinite future. Conventional green technologies have been applied in various fields. It will give solutions to exiting environmental issues. Materials and their handling can have a vast impact on the environment. It is very important to address the environmental and economic concerns in the production of new materials. The usage of renewable and sustainable materials has growing demands in society for pollution remediation and green products. The synthesis of some renewable and sustainable materials by different methods and their application in the development of green technology is presented in this chapter.

7.2   ROLE OF RENEWABLE AND SUSTAINABLE MATERIALS IN GREEN TECHNOLOGY

Because of the environment and sustainability problems, significant attainments in green technology through renewable and sustainable materials are occurred in recent times. Applications of various renewable and sustainable materials in green technology are described below:

7.2.1   BIOFUELS AS AN ALTERNATIVE ENERGY SOURCE

Energy demand is one of the major challenges of the twenty-first century. The studies show that the world's energy demand will keep growing and reach up to 50% and more by 2030 in all areas as a result of escalating expansion in population and fast technological development.[1] In order to meet this trend, sustainable and renewable energy sources are very important. The use of fossil fuels creates environmental issues through greenhouse gas emissions and subsequent global warming. Hence, the exploration of "clean" energy has to turn into one of the most overwhelming challenges. As a result, numerous substitute sources of energy like solar energy, wind, geothermal, hydroelectric, and biofuels are being examined and executed. Among these different potential sources of energy, biofuels are seen as actual means of attaining the aim of displacing fossil fuels in the short term.[2] Also, the fast hiking oil prize and emerging economic developments increase the demand for energy which resulted in the innovation of biofuels. Biofuels consist of energy derived from living organisms. They can be produced from starch, animal fats, vegetable oils, algal biomasses, or microbes.[3] All these sources are nonhazardous, eco-friendly, and can be restocked. The use of nonrenewable and nonsustainable fossil fuels can be replaced by biofuels. Ethanol, biodiesel, and biojet fuel are important types of biofuels. The wood and charcoal are the first used biofuel. The bioethanol is obtained from anaerobic fermentation of sugars by yeast while alcoholysis of vegetable or animal oils/fats gives biodiesel. Biofuels are more environmental friendly sources of energy because the burning of biofuels (alcohols, hydrogen) generates much lower (if any) carbon emission to the atmosphere contrast to the burning of fossil fuels.

On the basis of feedstock used and accessibility, biofuels can be divided into four categories. They are first, second, third, and fourth generations. The biofuel obtained from sugars and vegetable oils are included in the first generation. Second-generation biofuels are obtained from biomasses, generally from agricultural waste. The third generation is based on algae. Algae contain carbohydrates and lipids (fats) that can be converted into bioethanol and biodiesel, respectively. Biofuels are produced by a synthetic photosynthetic process using microorganisms that are included in the fourth generation. The largest quantity of biofuels produced in the first generation.[4] The use of biofuels reduces the harmful emission of CO_2, hydrocarbons, particulate matter, and SOx. This resulted in a reduction in greenhouse effects.[5]

The biofuels are produced by using food crops, thus it created a food versus fuel controversy. To solve these issue nonedible crops like Jatropha is used. This plant can live in a barren environment. The triglycerides found in the seeds can convert into biodiesel. But this plant needs a lot of water and nutrients to thrive. Some of the advantages and disadvantages of the usage of food crops for the production of biofuels are given below:

Advantages

- Plant crops are renewable since it can be planted many times
- It reduces the usage of nonrenewable source like petroleum
- It lowers the emission of greenhouse gases and pollution as compared to fossil fuels
- Cost-effective

Disadvantages

- The plating of crops requires a lot of water and fertilizers. The uses of fertilizers pollute the water bodies and groundwater system. In some areas, the availability of water is very low.
- The biofuels pollute the air as they ultimately end up with smoke.
- The fertility of the soil is increased by crop rotation. But the monoculture reduces soil fertility.
- The panting of the crop for biofuels reduces the land for food production

Shuba and Kifle reviewed the synthesis of biofuel using microalgae as promising sustainable and renewable materials for energy supply. They are emphasizing the advantages and different forms of these biofuels, their latest progress, and fabrication through genetic and metabolic engineering as well as prospect and promises and challenges in this industry.[6] Biofuels can also be prepared by biological methods using microbes. Elshahed gives a summary of the microbiological aspect of biofuel production along with state of the art, economic feasibility, and future direction of this area as renewable energy sources.[7] The production of biofuels by biological routes using bacterial and algae was discussed by Tajarudin et al.[8]

From all these reports we can say that the invention of biofuels resulted in a number of advantages like the reduction of greenhouse gas and it can replace fossil fuel.

7.2.2 ACTIVATED CARBON FROM BIOMASS FOR WATER TREATMENT

The activated carbon can be prepared by using bituminous coal, coke, and coconut shells. A large number of agricultural bi-products are also been used as the precursor for the preparation of activated carbons. It includes the agricultural wastes like coir pith, banana pith, maize comb, rice straw, rice hulls, fruit stones, nutshells, pinewood, bamboo, sawdust, rice husk, olive stones, bagasse, sago waste, cassava peel, etc.[9–14] These materials increase the amount of CO_2 in the atmosphere and are lignocellulosic. Hence, they are excellent materials for the synthesis of activated carbon.

Activated carbon can be synthesized mainly by two processes: physical activation involving carbonization (or pyrolysis) of the lignocellulosic material followed by an activation step by an oxidizing gas such as air, carbon dioxide, steam or their mixtures; and chemical activation involving a single carbonization step of the precursor in the presence of a chemical agent (KOH, $NaOH$, H_3PO_4, $ZnCl_2$, H_2SO_4, $(NH_4)_2SO_4$, HCl, $MgCl_2$, HNO_3, or $CaCl_2$) followed by heating under a nitrogen flow at 450–900°C. There are a lot of researches done to develop an alternative synthesis method with time, energy, and chemical savings as generating high worth materials from renewable resources such as biomass waste. Hence, nonconventional methods like microwave heating and hydrothermal carbonization treatment are being introduced for activated carbon preparation. Thus, activated carbon is a processed carbon with very small macro/micropores with a high surface area for adsorption.

Dyes containing wastewaters are generated in several industries like paint, dyeing, textiles, and paper. Almost nearly 40,000 dyes and pigments with over 7000 different chemical structures. The majority of them are not biodegradable. In each year, almost 10,000 various commercial dyes and pigments are produced worldwide. During the dyeing process, manufacturing, and processing operations, 20–27% of the dye is lost in the effluent and approximately 20% of this lost dye enters the industrial wastewaters. Dyes containing wastewaters required to treat appropriately because these contaminants may have carcinogenic, teratogenic, and mutagenic effects on both humans and aquatic life. Hence, the treatment of these dyes contaminated effluents with cost-effective and efficient technologies such as adsorption on activated carbons is of high scientific and public interest.[15–17] The various studies on the literature show the application of activated carbon in the removal of pollutant dyes like methylene blue,[18,19] acid blue,[20] Direct Blue 2B, Direct Green B,[21] Disperse Red 167,[22] Safranin O, Malachite Green,[23] etc. The activated carbon is also used for the treatment of water contaminated with a chlorinated compound like

2,4-dichlorophenoxyacetic acid,[24] polychlorinated biphenyls, chlordecone, hexachlorocyclohexane,[25] etc.

7.2.3 POLYMERIC MEMBRANE-BASED HEAT EXCHANGERS

Heat exchangers are the heart of the energy recovery system which is frequently used to recover waste energy either for industrial or building applications. Heat exchangers are also known as energy recovery ventilators. They transfer energy between the air exhausted from building and the outdoor supply air to reduce the energy consumption associated with the conditioning of ventilation air. The use of energy recovery ventilators reduces the heating energy expenditure appreciably.[26] The temperature difference between indoor and outdoor air affects heat recovery efficiency. The bigger the temperature difference is the higher the efficiency.

In the past, heat exchangers were originally made of metal to transfer sensible heat and polymers have been used only as the coating layer to prevent corrosion. But with the demand of using more sustainable and renewable materials and the needs of transferring latent heat, their surface area has been replaced with polymers and their composites. Yhaya gives a detailed review of polymeric membrane-based heat exchangers with a vast description of the effect of chemistry and chemical composition on their performance.[27] The main benefit of the polymeric membrane-based heat exchangers is that they can be used to recover energy in the hot–humid area in which they are placed to be integrated with the air conditioning system for energy recovery applications, energy based on heat and mass transfer mechanisms.[28] The development of these heat exchangers consists of porous fibers impregnated with a polymer-based solution that are able to capture moisture from the hot and humid air stream. In this system, heat is transferred and moisture is absorbed by the membrane surfaces of the heat exchanger, and hence, the temperature is decreased as much as possible which often depends on its ability and performance factors before entering the air conditioning system. The humid and hotter air is blown from inside, carrying the heat and moisture, and drying up the module as it is blown out to the environment. Therefore, the air conditioning system workload is reduced and so does the energy consumption.[29]

7.2.4 RENEWABLE SELF-HEALING MATERIALS FOR STRUCTURAL APPLICATION

In building structures, the crack formations in concrete structures are common because of its moderately lower tensile strength and action of different load and nonload factors. There are different reasons behind this cracking; some of them are plastic shrinkage, thermal stresses, drying shrinkage, rebar corrosion, external loading, and or coupled effect of multiple factors. Cracks can be manually repaired but there are several issues related to this repair operation, for example, impact on the environment, accessibility, and price. Self-healing is a promising solution to reduce manual intervention.

Self-healing materials are substances that are artificial or synthetically generated with the built-in capacity to involuntarily repair injure to themselves lacking any exterior analysis or human participation. The design of self-healing material gained attracted attention in various buildings and infrastructure applications.[30] Though, the majority of this application is incorporating an expansive and nonrenewable element in the structural form like concrete which starts to expand and fill voids and cracks when triggered by carbonation or moisture ingress.[31] Zulfiqar et al. reported the fabrication of superhydrophobic surfaces on building materials, which has the added quality to reinstate its properties after rigorous scratch. These superhydrophobic surfaces were constructed from hydrophobic silica nanoparticles and commercially obtainable spray adhesive on three commercially obtainable construction materials, that is, bricks, marble, and glass. These superhydrophobic surfaces were capable to maintain the impact of sand particles traveling at a speed of 11.26 km/h, and also restore their superhydrophobic character by simple acetone treatment upon getting rigorous damages by knife scratches.[32]

The bio-based self-healing materials have gained numerous attention in recent years. Sustainable mechanism of self-healing by microbial-induced precipitation of calcium carbonate is studied recently to close and repair cracks. This self-healing of cracks by microbes involves precipitation of calcium carbonate by the direct action of bacteria species including *Bacillus subtilis* on calcium compounds like calcium lactate or by the disintegration of urea by ureolytic bacteria like *Bacillus sphaericus*. Calcium carbonate precipitation using microbes is well-suited with concrete and the procedure of formation is environmentally friendly. *Bacillus sphaericus* is identified as safe to human beings. Gupta et al. give a detailed review of the evaluation of crack healing by bacteria and the effectiveness of factors affecting bacterial self-healing.[33] The self-healing materials for building and infrastructure applications give significant benefits, as they would allow conquering the problems related to internal damage analysis and restoration.

7.2.5   BIOCOMPOSITES AS A BUILDING MATERIAL

Green or biocomposite materials include biopolymers and natural fibers from renewable resources, which reduce the elimination of nonrenewable waste, raw material usage, and decrease greenhouse gas productions. The construction industry uses a huge quantity of materials, the majority of which are originated from nonrenewable resources or resources that necessitate significant time to be renewed. Hence, biocomposite materials are introduced as building materials with a reduced impact on environmental and human health. The biocomposites are used in various construction applications. The use of natural fibers as building materials is an advantage to attain a sustainable construction. Yan and Chouw described the potential of sustainable construction with natural fibers by empathizing the usage of a composite column containing flax fiber reinforced polymer and coir fiber reinforced concrete (CFRC).[34] Another study reports the flexural properties of plain concrete and CFRC beams which are externally strengthened by flax fabric reinforced epoxy polymer composites.[35] CoDyre et al. reported the effect of foam core density on the behavior of sandwich panels with novel biocomposite unidirectional flax fibers-reinforced polymer skins, with a comparison to panels of conventional glass-FRP skins.[36] The improvement effect of alkali treatment on microstructure and mechanical properties of coir fiber reinforced-polymer composites and reinforced-cementitious composites as building materials are also reported.[37]

7.2.6   BIOPOLYMERS FOR IMPROVING SOIL PROPERTIES

Biopolymers are substances naturally produced by living organisms and are hence considered to be eco-friendly and sustainable. Among the biopolymers, the chitosan and cellulose have particular importance due to their numerous applications, large availability, and easiness for modification. They have a large number of environmental applications. These biopolymers have high coagulating and flocculating power and these properties are very helpful in wastewater clarification, and it reduces the dependability on synthetic polyelectrolytes. Biopolymer-based hydrogels and nanocomposite films act as efficient biosorbents for eliminating a large number of organic and inorganic pollutants from wastewater. Particularly, it can adsorb heavy metal and dye. It also has many environmental applications like antidesertification, natural biosealants for preventing concrete leaks, and proton conducting membranes in electrochemical devices.

The biopolymers are also used for the enhancement of soil behavior.[38] Biopolymers have long been identified as viable soil conditioners because they stabilize soil surface structure and pore continuity. Acid-hydrolyzed cellulose microfibrils can be used for soil stabilization. The productivity of the soils can be improved by using a low concentration of biopolymers from plant fibers to augment water holding capacity improves the physical properties of soils by binding soils particles together reducing the losses of water by evaporation and deep percolation.[39] Soil erosion decreases water quality and agricultural productivity through the loss of valuable "topsoil" and runoff of agricultural chemicals. In a study, a series of biopolymers added to irrigation water were tested to reduce erosion-induce soil losses. Different polymer additives in irrigation water were tested for their efficacy in reducing soil losses during irrigation. Starch xanthate, cellulose xanthate, and acid hydrolyzed cellulose microfibrils, all appear promising for reducing soil runoff.[40]

Hataf et al. reported the potential of clay soil stabilization by a biocompatible chitosan solution that is synthesized from shrimp shell waste. The chitosan solution is used in different concentrations to evaluate its potentials on the mechanical properties of clay soil at different curing times and conditions. The results showed that the incorporation of chitosan has the potential to increase the interparticle interaction of the soil particles which leads to improved mechanical properties.[41] In another study, protein-based biopolymers (casein and sodium caseinate salt) were used for soil strengthening.[42] The thermo-gelatin polymers were also used for soil strengthening.[43]

7.2.7 PLANT FIBERS-BASED BIOCOMPOSITES AS SUSTAINABLE AND RENEWABLE GREEN MATERIALS

Plant fibers are a renewable resource and are an important group of reinforcing materials. It attained heavy attention in sustainable technology because of its plentiful occurrence and ease of access. Also, the composite materials of plant fibers are environment friendly, lightweight, and with high particular properties. Plant fibers-reinforced composites are made up of plant fibers and biodegradable polymer, as a matrix. In recent times, the plant fibers-reinforced composites replaced conventional fibers-reinforced composites, particularly glass fibers-reinforced composites. It is predicted that by 2020 fibers derived from bio-based resources will represent up to 28% of the whole market of reinforcement materials.[44] Owing to their peculiar properties like high mechanical strength, exceptional biocompatibility, they have gained extra interest and developing fields in materials technology. Plant fibers-reinforced composites have different industrial applications like internal parts of automotive and building structures and are used as a filler material. It is also widely used in the packaging, construction, transportation, and aviation industry.[45]

7.2.8 BIO-BASED EPOXY THERMOSETS FOR ENGINEERING APPLICATIONS

Epoxies are one of the most commonly used engineering thermosets because of their excellent properties like high rigidity, easy to process, greater tensile strength, good chemical as well as thermal resistance, outstanding electrical strength, exceptional solvent resistance, and compositional versatility. Though, these epoxy thermosets have some serious drawbacks like low toughness, high brittleness, and comparatively high cost.[46] In order to overcome these limitations, branch generating moiety has been introduced to the epoxy matrix. The presence of large numbers of reactive end functional groups in hyperbranched epoxy resins gives strength and toughness to the epoxy thermosets.[47]

In order to develop sustainable materials, renewable bio-based feedstocks like cardanol, tannin, lignin, glucose, vegetable oil, etc are used for the modification of epoxy resins. It gives biodegradability to the thermosets.[48] Vegetable oils have attained extensive interest because of its benefits like nontoxicity, renewability, effortless modification, biodegradability, and environment-friendly nature. Among the various types of vegetable oils, castor oil has been extensively used in industries due to its easy availability in large quantities and exceptional chemical composition.[49] De et al. prepared biodegradable hyperbranched epoxy from the polyester polyol of the monoglyceride of castor oil which exhibited outstanding adhesive strength, chemical resistance, good impact strength, scratch hardness, and biodegradability.[50] Das and Karak reported the synthesis of epoxy resin using *Mesua ferera* seed oil to give flexibility to the prepared epoxy thermoset.[51]

7.2.9   LACTIC ACID AS ENVIRONMENTAL-FRIENDLY CHEMICAL RESOURCES

Inventions of versatile chemical resources from renewable and sustainable materials have attained significant attraction in recent years. In this regard, the production of lactic acid from bacteria is seen as an option to cope with environmental problems and is cost-effective. Lactic acid (2-hydroxy propionic acid or 2-hydroxypropanoic acid) is an important organic acid with molecular formula $CH_3CHOHCOOH$. It is a chiral molecule, which exists as enantiomers L- and D-lactic acid. The lactic acid can be synthesized by using chemical synthesis and by microbial fermentation. By chemical synthesis, a racemic mixture of D- and L-lactic acid is obtained. But in the microbial fermentation process, an optically pure L(+)- or D(-)-lactic acid is obtained.[52] Also, this method has many advantages compared to chemical synthesis. Various inexpensive materials like molasses and other residues from agriculture and agro-industry have been used as substrates for lactic acid fermentation. Moreover, the efficiency of microorganisms for lactic acid synthesis can be enhanced by gene modification.[53,54] The lactic acid is extensively used in the pharmaceutical, cosmetic, food, and chemical industries. Lactic acid is regarded as one of the most important hydroxycarboxylic acids because of its versatile applications as a flavoring, inhibitor of bacteria, and acidulant. Lactic acid can be easily converted into potentially useful chemicals such as various acids, esters, and biosolvents since it contains both carboxyl and hydroxyl groups.[55] Lactic acid is also used as a feedstock monomer for the production of biodegradable poly-L-lactic acid, a superior substitute for synthetic polymers derived from petroleum resources.[56,57] In the sono-chemical synthesis of pyrrole derivatives, lactic acid acts as a bio-based green solvent. Andreev et al. stated that lactic acid fermentation helped to decrease the number of pathogens and to reduce the nutrient loss and hence increasing the agricultural value of plants.[58] Thus, the demand for lactic acid has been increasing significantly because of its various promising applications.

7.2.10   NATURAL FIBERS FOR BUILDING THERMAL INSULATION

The energy consumption of a building is strongly dependent on the characteristics of its envelope. The thermal performance of external walls is one of the important factors to increase the energy efficiency of the construction sector and to decrease greenhouse gas emissions. Thermal insulation is one of the best ways to reduce energy consumption. Thermal insulation systems and materials intend at reducing the transmission of heat flow. Building insulation is usually done with materials obtained from petrochemicals like polystyrene or from natural sources processed with high energy consumptions. These materials cause considerable harmful effects on the environment mostly due to the consumption of nonrenewable materials and fossil energy and troubles in reusing or recycling the products at the end of their lives. Hence, the researchers developed thermal and acoustic insulating materials using natural or recycled materials.
Aditya et al. reported the current progress on the building thermal insulation and also analyzed the life-cycle analysis and potential emissions reduction by using appropriate insulation materials.[59] A review of the important commercialized insulation materials is given by Schiavoni and coworkers.[60] The agro wastes like cereal straw, hemp, and olive waste were also used for the preparation of building insulation materials.[61]

7.3   CONCLUSIONS

This chapter briefly discusses the development of green technology through renewable and sustainable materials. The development of high-performance materials using renewable and sustainable materials is increasing worldwide and remarkable achievement occurs in green technology through these materials.

KEYWORDS

- **green technology**
- **renewable materials**
- **environmental**
- **sustainable materials**
- **cost-effective**

REFERENCES

1. Mardiana, A.; Riffat, S. B. Building Energy Consumption and Carbon Dioxide Emissions: Threat to Climate Change. *J. Earth Sci. Clim. Change* **2015,** *S3*, 1.
2. Maness, P. C.; et al. Photobiological Hydrogen Production-Prospects and Challenges. *Microbe* **2009,** *4* (6), 659–667.
3. Chisti, Y. Biodiesel from Microalgae. *Biotechnol. Adv.* **2007,** *25* (3), 294–306.
4. Song, D.; Fu, J.; Shi, D. Exploitation of Oil-Bearing Microalgae for Biodiesel. *Chin. J. Biotechnol.* **2008,** *24* (3), 341–348.
5. Bringezu, S. Ed. *Towards Sustainable Production and Use of Resources: Assessing Biofuels*. UNEP/Earthprint, 2009.
6. Popp, J.; et al. The Effect of Bioenergy Expansion: Food, Energy, and Environment. *Renewable Sustainable Energy Rev.* **2014,** *32*, 559–578.
7. Shuba, E. S.; Kifle, D. Microalgae to Biofuels: 'Promising' Alternative and Renewable Energy, Review. *Renewable Sustainable Energy Rev.* **2018,** *81*, 743–755.
8. Elshahed, M. S. Microbiological Aspects of Biofuel Production: Current Status and Future Directions. *J. Adv. Res.* **2010,** *1* (2), 103–111.
9. Tajarudin, H. A.; et al. Energy Recovery by Biological Process. *Renewable Energy and Sustainable Technologies for Building and Environmental Applications*; Springer: Cham, 2016; pp 227–249.
10. Ahmedna, M.; Marshall, W. E.; Rao, R. M. Production of Granular Activated Carbons from Select Agricultural By-Products and Evaluation of Their Physical, Chemical and Adsorption Properties. *Bioresour. Technol.* **2000,** *71* (2), 113–123.
11. Namasivayam, C.; Kadirvelu, K. Activated Carbons Prepared from Coir Pith by Physical and Chemical Activation Methods. *Bioresour. Technol.* **1997,** *62* (3), 123–127.
12. Chafidz, A.; et al. Preparation of Activated Carbon from Banana Peel Waste for Reducing Air Pollutant from Motorcycle Muffler. MATEC Web of Conferences. Vol. 154. EDP Sciences, 2018.

13. Foo, K. Y.; Hameed, B. H. Preparation and Characterization of Activated Carbon from Pistachio Nut Shells via Microwave-Induced Chemical Activation. *Biomass Bioenergy* **2011,** *35* (7), 3257–3261.

14. Le Van, K.; Thi, T. T. L. Activated Carbon Derived from Rice Husk by NaOH Activation and Its Application in Supercapacitor. *Prog. Nat. Sci.* **2014,** *24* 3, 191–198.

15. Moreno-Piraján, J. C.; Giraldo, L. Adsorption of Copper from Aqueous Solution by Activated Carbons Obtained by Pyrolysis of Cassava Peel. *J. Anal. Appl. Pyrolysis* **2010,** *87* (2), 188–193.

16. Demirbas, A. Agricultural Based Activated Carbons for the Removal of Dyes from Aqueous Solutions: A Review. *J. Hazard. Mater.* **2009,** *167* (1–3), 1–9.

17. Feng, Y.; et al. Methylene Blue Adsorption onto Swede Rape Straw (Brassica napus L.) Modified by Tartaric Acid: Equilibrium, Kinetic and Adsorption Mechanisms. *Bioresour. Technol.* **2012,** *125,* 138–144.

18. Mitter, E. K.; et al. Analysis of Acid Alizarin Violet N Dye Removal Using Sugarcane Bagasse as Adsorbent. *Water Air Soil Pollut.* **2012,** *223* (2), 765–770.

19. Benadjemia, M.; et al. Preparation, Characterization and Methylene Blue Adsorption of Phosphoric Acid Activated Carbons from Globe Artichoke Leaves. *Fuel Process. Technol.* **2011,** *92* (6), 1203–1212.

20. Foo, K. Y.; Hameed, B. H. Preparation of Activated Carbon from Date Stones by Microwave Induced Chemical Activation: Application for Methylene Blue Adsorption. *Chem. Eng. J.* **2011,** *170* (1), 338–341.

21. Valix, M.; Cheung, W. H.; McKay, G. Roles of the Textural and Surface Chemical Properties of Activated Carbon in the Adsorption of Acid Blue Dye. *Langmuir* **2006,** *22* (10), 4574–4582.

22. Malik, P. K. Use of Activated Carbons Prepared from Sawdust and Rice-Husk for Adsorption of Acid Dyes: A Case Study of Acid Yellow 36. *Dyes Pigm.* **2003,** *56* (3), 239–249.

23. Wang, L. Removal of Disperse Red Dye by Bamboo-Based Activated Carbon: Optimisation, Kinetics and Equilibrium. *Environ. Sci. Pollut. Res.* **2013,** *20* (7), 4635–4646.

24. Salima, A.; et al. Application of Ulva lactuca and Systoceira stricta Algae-Based Activated Carbons to Hazardous Cationic Dyes Removal from Industrial Effluents. *Water Res.* **2013,** *47* (10), 3375–3388.

25. Hameed, B. H.; Salman, J. M.; Ahmad, A. L. Adsorption Isotherm and Kinetic Modeling of 2, 4-D Pesticide on Activated Carbon Derived from Date Stones. *J. Hazard. Mater.* **2009,** *163* (1), 121–126.

26. Amstaetter, K.; Eek, E.; Cornelissen, G. Sorption of PAHs and PCBs to Activated Carbon: Coal Versus Biomass-Based Quality. *Chemosphere* **2012,** *87* (5), 573–578.

27. Rasouli, M.; Simonson, C. J.; Besant, R. W. Applicability and Optimum Control Strategy of Energy Recovery Ventilators in Different Climatic Conditions. *Energy Build.* **2010,** *42* (9), 1376–1385.

28. Yhaya, M. F. Polymeric Heat Exchangers: Effect of Chemistry and Chemical Composition to their Performance. *Renewable Energy and Sustainable Technologies for Building and Environmental Applications*; Springer: Cham, 2016; pp 51–67.

29. Ahmad, M. I.; Mansur, F. Z.; Riffat, S. Applications of Air-to-Air Energy Recovery in Various Climatic Conditions: Towards Reducing Energy Consumption in Buildings.

Renewable Energy and Sustainable Technologies for Building and Environmental Applications; Springer: Cham, 2016; pp 107–116.

30. Yang, P.; et al. Testing for Energy Recovery Ventilators and Energy Saving Analysis with Air-Conditioning Systems. *Procedia Eng.* **2015**, *121*, 438–445.

31. Qian, S.; et al. Self-Healing Behavior of Strain Hardening Cementitious Composites Incorporating Local Waste Materials. *Cem. Concr. Compos.* **2009**, *31* (9), 613–621.

32. Sisomphon, K.; Copuroglu, O.; Fraaij, A. Application of Encapsulated Lightweight Aggregate Impregnated with Sodium Monofluorophosphate as a Self-Healing Agent in Blast Furnace Slag Mortar. *Heron* **2011**, *56* (1/2), 1–20.

33. Zulfiqar, U.; et al. Durable and Self-Healing Superhydrophobic Surfaces for Building Materials. *Mater. Lett.* **2017**, *192*, 56–59.

34. Gupta, S.; Pang, S. D.; Kua, H. W. Autonomous Healing in Concrete by Bio-Based Healing Agents–A Review. *Constr. Build. Mater.* **2017**, *146*, 419–428.

35. Yan, L.; Chouw, N. Experimental Study of Flax FRP Tube Encased Coir Fibres Reinforced Concrete Composite Column. *Constr. Build. Mater.* **2013**, *40*, 1118–1127.

36. Yan, L.; Su, S.; Chouw, N. Microstructure, Flexural Properties and Durability of Coir Fibres Reinforced Concrete Beams Externally Strengthened with Flax FRP Composites. *Compos. Part B* **2015**, *80*, 343–354.

37. CoDyre, L.; Mak, K.; Fam, A. Flexural and Axial Behaviour of Sandwich Panels with Bio-Based Flax Fibres-Reinforced Polymer Skins and Various Foam Core Densities. *J. Sandwich Struct. Mater.* **2018**, *20* (5), 595–616.

38. Yan, Libo; et al. Effect of Alkali Treatment on Microstructure and Mechanical Properties of Coir fibres, Coir Fibres Reinforced-Polymer Composites and Reinforced-Cementitious Composites. *Constr. Build. Mater.* **2016**, *112*, 168–182.

39. Ayeldeen, M.; et al. Enhancing Mechanical Behaviors of Collapsible Soil Using Two Biopolymers. *J. Rock Mech. Geotech. Eng.* **2017**, *9* (2), 329–339.

40. Maghchiche, A.; Haouam, A.; Immirzi, B. Use of Polymers and Biopolymers for Water Retaining and Soil Stabilization in Arid and Semiarid Regions. *J. Taibah Univ. Sci.* **2010**, *4* (1), 9–16.

41. Orts, W. J.; Sojka, R. E.; Glenn, G. M. Biopolymer Additives to Reduce Erosion-Induced Soil Losses During Irrigation. *Ind. Crops Prod.* **2000**, *11* (1), 19–29.

42. Hataf, N.; Ghadir, P.; Ranjbar, N. Investigation of Soil Stabilization Using Chitosan Biopolymer. *J. Cleaner Prod.* **2018**, *170*, 1493–1500.

43. Fatehi, H., et al. A Novel Study on Using Protein Based Biopolymers in Soil Strengthening. *Constr. Build. Mater.* **2018**, *167*, 813–821.

44. Chang, I.; et al. Soil Strengthening Using Thermo-Gelation Biopolymers. *Constr. Build. Mater.* **2015**, *77*, 430–438.

45. Shah, D. U. Developing Plant Fibres Composites for Structural Applications by Optimising Composite Parameters: A Critical Review. *J. Mater. Sci.* **2013**, *48* (18), 6083–6107.

46. Ramesh, M.; Palanikumar, K.; Hemachandra Reddy, K. Plant Fibres Based Bio-Composites: Sustainable and Renewable Green Materials. *Renewable Sustainable Energy Rev.* **2017**, *79*, 558–584.

47. May, C. A. Introduction to Epoxy Resins. *Epoxy Resins Chem. Techno.* **1988**, *2*, 1–8.

48. De, B.; et al. Biodegradable Hyperbranched Epoxy from Castor Oil-Based Hyperbranched Polyester Polyol. *ACS Sustainable Chem. Eng.* **2013**, *2* (3), 445–453.

49. Auvergne, R.; et al. Biobased Thermosetting Epoxy: Present and Future. *Chem. Rev.* **2013,** *114* (2), 1082–1115.

50. Gupta, A. P.; Ahmad, S.; Dev, A. Modification of Novel Bio-Based Resin-Epoxidized Soybean Oil by Conventional Epoxy Resin. *Polym. Eng. Sci.* **2011,** *51* (6), 1087–1091.

51. De, B.; et al. Biodegradable Hyperbranched Epoxy from Castor Oil-Based Hyperbranched Polyester Polyol. *ACS Sustainable Chem. Eng.* **2013,** *2* (3), 445–453.

52. Das, G.; Karak, N. Epoxidized *Mesua ferrea L.* Seed Oil-Based Reactive Diluent for BPA Epoxy Resin and Their Green Nanocomposites. *Progr. Org. Coat.* **2009,** *66* (1), 59–64.

53. Boonpan, A.; et al. Separation of D, L-lactic Acid by Filtration Process. *Energy Procedia* **2013,** *34,* 898–904.

54. Okano, K.; et al. Biotechnological Production of Enantiomeric Pure Lactic Acid from Renewable Resources: Recent Achievements, Perspectives, and Limits. *Appl. Microbiol. Biotechnol.* **2010,** *85* (3), 413–423.

55. Datta, R.; Henry, M. Lactic Acid: Recent Advances in Products, Processes and Technologies-A Review. *J. Chem. Technol. Biotechnol. Int. Res. Process Environ. Clean Technol.* **2006,** *81* (7), 1119–1129.

56. Gao, C.; Ma, C.; Xu, P. Biotechnological Routes Based on Lactic Acid Production from Biomass. *Biotechnol. Adv.* **2011,** *29* (6), 930–939.

57. Khunnonkwao, P.; et al. Purification of L-(+)-Lactic Acid from Pre-Treated Fermentation Broth Using Vapor Permeation-Assisted Esterification. *Process Biochem.* **2012,** *47* (12), 1948–1956.

58. Wang, Y.; Tashiro, Y.; Sonomoto, K. Fermentative Production of Lactic Acid from Renewable Materials: Recent Achievements, Prospects, and Limits. *J. Biosci. Bioeng.* **2015,** *119* (1), 10–18.

59. Andreev, N. Lactic Acid Fermentation of Human Excreta for Agricultural Application." *J. Environ. Manage.* **2018,** *206,* 890–900.

60. Aditya, L.; et al. A Review on Insulation Materials for Energy Conservation in Buildings. *Renewable Sustainable Energy Rev.* **2017,** *73,* 1352–1365.

61. Schiavoni, S.; Bianchi, F.; Asdrubali, F. Insulation Materials for the Building Sector: A Review and Comparative Analysis. *Renewable Sustainable Energy Rev.* **2016,** *62,* 988–1011.

62. Liuzzi, S.; Sanarica, S.; Stefanizzi, P. Use of Agro-Wastes in Building Materials in the Mediterranean Area: A Review. *Energy Procedia* **2017,** *126,* 242–249.

CHAPTER 8

Green Technologies for Nutrient Cycles in Crop and Livestock Systems Management

SHRIKAANT KULKARNI*

Department of Chemical Engineering, Vishwakarma Institute of Technology, 666 Upper Indira Nagar, Bibwewadi, Pune 411037, India

Corresponding author. E-mail: shrikaant.kulkarni@vit.edu

ABSTRACT

The nutrient cycle is made up of two important geochemical fluxes, namely, nitrogen and phosphorous. The nutrient cycle asks for timely action on the environmental policy front in terms of soil management, farming systems, sewage treatment, and so on. The nutrient cycle unfortunately is not getting the necessary attention it deserves as the nutrients policy is vulnerable. Nutrients such as nitrogen (N) and phosphorus (P) are essential for the growth and development of organisms. Ecosystems are responsible for controlling the flows and concentration levels of nutrients through a host of complex processes including biodiversity. Nutrients cycles have so far been modified to a substantial extent by human intervention mainly agriculture over the time, with its own consequences for not only a range of ecosystems but also to human wellbeing. The nutrients absorption and retention capacity of terrestrial ecosystems supplied by way of fertilizers or deposition have been depleted because of many large ecosystems into large scale but low diversity agricultural systems. Further with the reduction in the buffering capacity of ecosystems like riparian forests, wetlands, and estuaries, the nutrients in excess leach into groundwater, rivers, and lakes and are subsequently transported to coastal ecosystems. Although agriculture intensification is the only option to meet the future food demands and to check conversion of land from natural vegetation to agriculture, excessive flows of N have contributed to eutrophication, acidification of freshwater bodies, and coastal marine ecosystems. Nitrogen losses promote global warming and, to a certain extent, instrumental in creating ground-level ozone and depletion of stratospheric ozone layer.

This chapter is an overview of the sources of N losses at the different stages of N cycling in agro-ecosystems, technical solutions to cut down the quantum of N losses, and throw light on the analysis supported by integrated tools and indicators in two contrasting situations: a low-input and a high-input system from the developing and an industrial world, respectively. Moreover, the research needs for better assessment of the expected benefits to be achieved at a global level in land productivity and erosion in environmental impact by improving nutrient cycling management discussed too.

8.1 INTRODUCTION

Nutrients like nitrogen (N) and phosphorus (P) are essential for the growth and development of all organisms. Ecosystems maintain a balance of the flows and concentrations of nutrients using many complex processes even species diversity. Nutrient cycles are largely modified by human intervention with agriculture in preference over the centuries,[1–2] with consequences for a plethora of ecosystems, and for wellbeing of humanity.[3] The inbuilt capacity of terrestrial ecosystems to absorb and retain the nutrient contents obtained from either fertilizers or depositions has been constrained by the large number of low-diversity ecosystems.[4] With the reduction in their buffering capacity like riparian forests, wetlands, and estuaries, surplus nutrients make their way into groundwater, rivers, and lakes and are conveyed to coastal ecosystems.[5]

In preindustrial periods, the annual flux of N from the atmosphere to the land and aquatic ecosystems was approximately 110–210 MT N year^{-1}. Anthropological activities contributed another 170 MT N year^{-1} doubling the pace of generation of reactive N on Earth's surface.[6–7] Phosphorus also accumulates in ecosystems at a velocity of 10.5–15.5 MT P year^{-1}, against preindustrial pace of 1–6 MT P year^{-1}, due to the use of phosphorus-based fertilizers for agricultural purposes.

Most of these accumulate in soils.[8] Nutrient contents on agricultural land allowed a marked rise in food production, particularly in industrial world[9] but at the expense of rise in emissions of green house gases (GHGs) and contamination of freshwater and coastal ecosystems as against the nutrient excessive supply in the developed countries, leaving vast areas of the earth particularly in Africa and Latin America, where harvesting was done without supplying nutrients, externally had brought about land erosion and loss of soil fertility, with evil effect on human food and nutrition.[10–11]

The world's population is likely to have increased around 40% by 2050.[12] The demand for food will double as a result probably.[13] Huge food demand is expected in developing world particularly in sub-Saharan Africa and Latin America where the rates of population growth are projected as highest. The tropical and neotropical areas will have major expansions in cultivated land and the rise in production is likely to be huge.[14] Many studies reveal that agriculture intensification is the only option to meet the future food demand and to check transformation of land from natural vegetation to agricultural one.[15,16] Latest studies on projections of N fertilizer use show rise between 10% and 80% by 2020 based on future food demands and the measure adopted by farmers to check soil erosion.[17] Millennium Ecosystem Assessment portrays that the global flow of N to coastal ecosystems is bound to rise by 10–20% by 2030.[18] The N contents in rivers would remain stable in majority of industrial nations, while a 20–30% rise is likely for developing world. This is as a result of rise in N inputs to surface water because of increase in food production and N fertilizer, deposition, biological fixation, and animal manure in agricultural systems.[19] Subsequently, producing food for the world population, but with no profound effect world ecosystems, will be a global challenge in the reckoning without any universally acceptable solution.[20]

Managing nutrient cycles in agriculture (crop and livestock systems) can contribute in accomplishing the goal with certainty. Conserving and applying nutrients with improved efficiency can help lead to rise in crop yields and check ecological impact. Following are the questions worth studying:

- What could be obtained by nutrient cycles' management?
- What would be the meaningful, productive, and environmentally relevant goals and how can they be accomplished in reality?

N is one of the most limiting parameters in crop[21,22] and a determinant in animal production, with the help of proteins and amino acids supply from the feed.[23] Nitrogen exists in various forms in agro-ecosystems, with varied life span, and it is prone to different anthropogenic losses harmful to the environment.[24] Surplus N flows contribute to eutrophication and acidification of freshwater and marine ecosystems. Nitrogen losses fuel global warming effect and are also responsible to generate ground-level ozone, depletion of the stratospheric ozone layer.[25]

8.2   NITROGEN CYCLING AND MAJOR LOSSES

N Losses are in four major forms such as ammonia (NH_3), nitrate (NO_3^-), nitrous oxide (N_2O), nitrogen oxides (NOx), and nitrogen (N_2). The first three forms of losses are considered as the most harmful ones. International Panel on Climate Change (IPCC)[26,27] for regional and global inventories has presented ranges of emissions rates in the literature.

The three major steps for N losses are:

- during manure handling, manure collection in barns, storage, and processing
- during grazing after use and direct deposition
- through leaching and runoff after use and direct deposition on the surface of land.

8.3   NITROGEN EMISSIONS IN MANURE HANDLING

Lower N use efficiency almost 90% of N ingested by livestock is excreted as feces and urine although efficiency apparently changes a lot among various species from beef cattle (5%) to poultry (34%).[28] A global projection is about 10% for N use efficiency by livestock, and pastoral systems. Major chunk of N in excreta rapidly mineralizes, typically for livestock protein-rich food.[29,30] In urine, specifically greater than 70% of the N exists as urea and uric acid as nitrogen-rich prominent compounds in poultry discharges.[31] Fast decay of urea and uric acid to ammonium ion form brings about huge NH_3 losses by volatilization during the handling of manures. The emissions from manures are influenced by many factors, in particular, the concentration of ammonia, room temperature, humidity level, and the pH.[32] The amount of losses of ammonia determined from excreta of various species are as follows: sheep and goats—5–9%; dairy cattle—25–30%; camels, buffalo, and horses—16–18%; and pigs and poultry—35%. A projected ammonia liberation rate is of 20% of N at global level in manure.[33–35]

A very small amount of the total N excreted is transformed into N_2O during manure handling process.[36] However, the mineralization rate and subsequent NH_3 emissions are mainly dependent upon the manure composition, management, storage span, and temperature and are the key driving forces of N_2O emissions.[37] From manure mounds in Indonesia (Mali) losses were < 0.5% of initial N contents.[38] Under temperate conditions, N_2O emissions are abysmally low for manures in liquid form having no cover of natural crust.[39,40] N_2O emissions were determined varying from 0.25% to 1% for solid manures.[41] Larger emission rates of the order of 20% were noticed for thoroughly mixed deep bedding of cattle and pig.[42,43] The Emission rates of 1% and 2% for N_2O were determined for manure from sheep excretions, and cattle, poultry, and pigs, respectively.[44] Emission rates vary between 0.01% and 7% for slurries and pig waste, respectively.[45] Depending on current manure management in varied systems and regions of the world, in combination with IPCC emissions rates, Food and Agriculture Organization (FAO) proposed 1.5% as the emission rate for N_2O and in turn N at global level present in manure.[46]

8.4   NITROGEN EMISSIONS POSTDIRECT DEPOSITION AND USE

Temperate conditions bring about volatilization of NH_3 after use depending upon the type and quality of manure at the rates varying between 2–20% and 5–50% for solid manure and slurry, respectively.[47,48] A 20% N loss fraction for NH_3 volatilization, irrespective of whether it manure or direct deposition of excreta. It was determined that the N loss through ammonia volatilization from animal manure after use was 24.2% across the world.[49] The said losses were 15–20% at least.[50] Directly deposited excreta on exposure comparatively large N loss rates, in the form of ammonia, typically in tropical and dry environments.[51] Temperate conditions show emission rates varying between 1–7% of total N for cows grazing and 6–19% for monogastric herbivores.[52,53] The quality of grazed forages and environmental situations make N emissions from excreta on pastures not easy to determine because of variations in them. FAO suggested a volatilization rate of the order of 40% at the global level for direct deposition of excreta.[54] The standard IPCC rate of 20% is in harmony with that quoted in literature.

Losses on manure use and direct deposition on soil are considered as the highest N_2O emissions from livestock as a source.[55] The N_2O emissions are governed by many factors, such as spore space in soil filled with water, organic carbon contents, pH, temperature of soil, plant uptake, and rainfall patterns.[56] Emission rates varying between 0.2% and 4.6% were observed under temperate conditions.[57] They match for applied manure and directly deposited excreta. Emission computations are compared for tropical and temperate soils and livestock species. The present IPCC emission rate is 1% and ranges from 0.2–5% for directly deposited excreta, showing an uncertainty to a large extent.[58]

The N_2O loss rates are estimated at the global level to 0.5% and 2.1% for manure applied and direct deposition of excreta, respectively.[59] The amount of N emissions after applying mineral fertilizer is influenced by the type of fertilizer. It was determined that NH_3 emission losses from urea as 14% and 24%, respectively in tropical and temperate conditions.[60] The NH_3 loss from ammonium bicarbonate applied as a fertilizer is more of the order of 21% and 31% in tropical and temperate regions. For mineral fertilizer types employed across world, the NH_3 volatilizes at rate[61] of 14.5% and the N_2O liberates at 1% of the total N used.

8.5   NITROGEN LEACHING AND RUNOFF POSTDIRECT DEPOSITION AND USE

N loss takes place from soils and leaching is one of the mechanisms underlying it. Due to the comparatively poor N uptake efficiency of plants,[62] a better portion of the applied or deposited N is not absorbed by the plant tissues nor retained by the soil. It subsequently gets into the "Nitrogen cascade". N has mobility in the soil as nitrate in preference and gets leached easily to the groundwater. N in organic forms can also be leached into the water via runoff and lowers following heavy rains. N losses and pollution potential is influenced by soil and weather characteristics, intensity, frequency and grazing time span, and manure application rate.[63] Annual N leaching losses estimated from millet farms getting 8–10 t ha^{-1} of manure are about 90 kg N ha^{-1}. For example, in gardens of Niamey Niger, in urban areas, N leaching losses were determined to be 11.4 kg N ha^{-1}, for lands getting manure about 9 t ha^{-1}. Such losses estimated were 16–38 kg N ha^{-1} by installing lysimeters on a sandy soil cultivated with crops such as wheat and maize in Zimbabwe. The losses measured were 11–18 kg N ha^{-1} in control plot of lands cultivated with maize.[64] These losses were found to increase ranging between 15 and 49 kg N ha^{-1}, based on rainfall during the growing season (839 vs. 1396 mm) for fertilizer of amount 60 kg N ha^{-1}. Losses varied from 15 to 55 kg N ha^{-1} for fertilizer of amount 120 kg N ha^{-1}. Application of manure at the rates of 12.4 or 37.6 t ha^{-1} had a very little effect on N leaching. Losses varied between 4% and 12% of N applied in second and third seasons, while 25–41% in the 1st season with heavy rainfall. Comparatively more quantities of N are lost via leaching in agriculture of Africa fed, where fertilizer application is far less than already laid down 50 kg N ha^{-1}. N losses in runoff are normally below 5% of the fertilizer application rate.[65] In general N leaching and runoff from agricultural ecosystems to water, of N fertilizer, ranges between 11–39% and 24–79% for clay sandy soils, respectively. Globally 24% of the N used is lost and pollutes water bodies. As far as N directly deposited to areas grazed is concerned, the N losses rely upon the intensity of grazing. Stocking rates, % utilization of plants, and stubble heights are normally employed to explain grazing intensity. Moderate grazing intensity practices lead to almost no N losses.[66] However, intensive grazing ones increase N losses in leaching and runoff. N losses at global level following direct excreta deposition were found to be negligibly small in the FAO's global evaluation due to the large dominance of extensively grazed areas.

8.6   USE OF GREEN TECHNOLOGIES FOR CUTTING DOWN NITROGEN LOSSES

Following are the green technologies used to reduce N losses, controlling a host of vital processes in N cycle:

(1) cutting down N excretion in manure by altering livestock foods,
(2) enhance manure handling,
(3) regulate on-field N emissions,
(4) check on field leaching and runoff, and
(5) enhance N use efficiency by plants.

Each step can be sorted out by using selective technical initiatives.

8.7   CHECKING NITROGEN LEAKAGE FROM MANURE

A key route lies in increasing the animal N digesting efficiency (10% vs. 40% for crops). It can be done by breeding, animal health by stress reduction, and particularly using more balanced diets, that is, by optimizing protein intake in tune with the requirements of livestock[67] or altering diets to raise the distribution of excreta to fecal N. Improvement in feeding practices involve grouping of animals based on gender and phase of production, and enhancing the feed conversion ratio by modifying feed in tandem with physiological requirements. Reducing the inorganic N content in manure is amenable to reduction in N_2O and NH_3 emissions from stables while storing and following application to soil. For livestock, an appropriate balance in feed among degradable or otherwise proteins enhances nutrient absorption and has been found to check N excretion by 14–29% without influencing production quantities.[68] For swine, a lesser quantity of protein supplemented with synthetic amino acids reduces N excretion up to 29%, based on the diet composition in the beginning. Fiber and germ removal from corn is found to lower the concentration of N present in urine and the feces by 38%.[69] However, when best practices are employed, huge amounts of mineral N still are left behind in the manure and various technical options have to be explored at different stages to cut down N losses.

8.8   MANURE HANDLING IMPROVED METHODS

Housing of livestock and storage of manure cause contained N losses by using covered containers for the collection of manure, and properly sized storage facilities. In low-input systems, direct application or small storage time lowered nutrient losses and maintained the manure quality of organic fertilizers. Right storage capacity highly important to prohibit losses in the rainy season in places having a frequency of good enough rainfall is more. Application of an enclosed tank checks the dilution exercise and risk of overflow. Covering permanently with a lid lowers substantially by 81–91% of gaseous emissions such as NH_3 when stored. Covering with a natural crust on the slurry surface in a tank is quite efficient and effective.[70]

The option of N_2O gaseous emissions mitigation in storage is complex due to the trade-off between CH_4 and N_2O emissions. For example, switching over from straw- to slurry-based systems may lead to the reduction in N_2O but a rise in CH_4 emission. Different treatment technologies such as solid manure composting, slurry aeration are in place for the nutrients' concentration obtained from separated phases to derive products possessing more fertilizer value and easy transport and to lower local manure excesses and the pollution potential of water bodies via runoff and leaching.[71] Such technologies are used preferentially to control odor and to check hazardous gaseous emissions. However, many of such processes tend to more N losses in the gaseous NH_3 form. Use of organic substances to downsize the mineral pool of N has been recommended as an effective method to check gaseous emissions. Downsizing the substrate and preventing from maintaining aerobic conditions contain losses of NH_3 and might induce a bit of N_2O losses, although it depends more upon substrates available as energy sources.[72]

8.9   CHECKING IN-FIELD N GASEOUS EMISSIONS

Soil N emissions are largely governed by temperature and soil humidity and by agricultural practices too. N_2O emissions from the use of slurry to grassland were lowered on storing slurry for six months or passing it through an anaerobic digester such as for biogas production as there are low C contents in the slurry. However, such technical initiatives rise NH_3 volatilization. Rapid induction and shallow injection techniques for manure lower N loss by a minimum 50%, while deep injection technique in the soil removes this loss. Another technical solution for checking N emissions is the application of nitrification inhibitors (NIs). NIs can be employed on pastures to work on urinary N. The rate at which NH_3 liberates from mineral fertilizers can be checked from 14.0% to 4.5% when anhydrous ammonia is injected.[73] N_2O emissions rise in exponential terms with N in fertilizers, and the amount of losses relies upon soil, fertilizer type, drainage conditions, and fertilizer application trials to determine optimal crop response and economic rates, which should involve determination of gaseous emissions for estimating extent of environmental contamination.

8.10 CHECKING RUNOFF AND LEACHING

The key to lowering on-field N losses is the optimization of organic and mineral fertilizers use in respect of environmental conditions, like timing, quantities, and way it is applied in line with crop physiology and climate.[74] N application in tandem with the growth period demanding nutrients seems to be simple, but not so on field. For organic fertilizers, nutrient availability to plants alters depending upon animal foods and manure quality, management practices, and climatic conditions. Timespan between N application and mineralization can be very vital, particularly in both high-input temperate and low-input tropical environments for manures. Organic N retained by soils can mostly mineralize with poor N uptake by crops and is vulnerable to leaching, such as in early spring in temperate areas and with the start of the rainy season in tropical environments.[75] Spatial soil heterogeneity is high in the developing world, typically in mixed crop–livestock systems (CLIs).[76]

The existence and quality of the soil cover is equally vital land management factor responsible for the leaching intensity. For grazing, regulating grazing season, frequency, intensity, and distribution of livestock can increase vegetation cover and check leaching and runoff. In temperate areas, losses in excess from manure can be contained by preventing late fall and winter grazing. For cropping systems, a permanent vegetation cover, for example, zero tillage cultural systems, can lower N leaching substantially.[77]

8.11 PLANTS EFFICIENT USE OF NITROGEN

Enhancing N uptake efficiency of plants, like animals, means rise in crop yield/N input. It would lead to a lowering down of fertilizer applications and commensurate losses. It is possible through conventional technologies like irrigation, conservation of agriculture, genetic growth as transgenic plants, improved plantation techniques, drainage, and so on. N fixation by legumes has a key role in lowering down fertilizer application by introducing legumes in grasslands and inter-cropping. Optimized practices and environmental factors for cultivation of plants lead to an enhanced efficiency by 72%. Other 28% are intrinsic loss derived from ecosystems that are cultivated.[78]

8.12 USE OF INTEGRATED TOOLS AND INDICATORS FOR IMPROVING N CYCLING

Lowered N emissions in manure storage exercise can be made up by more N emissions after manure application. Subsequently, a comprehensive view of the N cycling in agro-ecosystems is required. Simulation can elaborate this holistic view. This is described using two model-based experiments carried out in two contrast situations. The first one in southern Africa, pertaining to low-input systems wherein soil fertility is reducing and the preference is to enhance farm yield to meet household food demands while the second one is related to a peripheral area of the European Union and is about high-input systems wherein the major problem is the lowering down N surpluses and potential risks of freshwater contamination.[79]

8.13 USE OF ORGANIC AND INORGANIC RESOURCES OF NITROGEN FOR PRODUCTIVE AND EFFICIENT USE OF NITROGEN

Southern Africa has poor sandy soils in vast areas. Conventionally, farmers maintain soil fertility by combining local organic resources, litter, livestock manure, and so on to produce food. However, rising human population has resulted in stiff competition for use of land and organic resources required to manage soil fertility. Most farmers in sub-Saharan Africa bring under cultivation small plots of land (<2 ha), use small amounts of mineral fertilizers, and own few or no livestock, which is responsible for the depletion of nutrients. Fertilizers are very much required to boost food production in Africa. However, it is not rendered much effective due to soil degradation thereby ending up small amount of food production.[80] Soil heterogeneity, accompanied by management, leads to comparatively small areas used in the croplands by farmers to produce food. To increase land productivity, the availability of fertilizers or manures, and organic inputs, which in Africa exceed farm scale have to be explored. The modeling system uses results obtained from experimentation from a community farming system in Zimbabwe to understand potential to give an impetus to food production, self-sufficiency, and N use efficiency, and likely consequences on the environment.

8.14 METHODOLOGY USED

Nuances farmism is a model applicable to farm scale which connects simulation findings derived using four submodels. The combination of these submodels provides for representation of short and long-term feedbacks pertaining to correlation between crops and livestock, and their implications on the natural resources of farms. Crop and soil models are coupled at farm scale in the model FIELD (farm-scale resource interactions, use efficiencies, and long-term soil fertility development).[81] Couplings between crops and soils can be simulated for various types of fields. LIVSIM (LIVestock SIMulator) simulates production of animals depending upon genetic potential, availability, and quality of feed. The nutrients dynamism via manure collection and storage can be simulated by Manure Heap Simulation Model (HEAPSIM) determining mass and nutrient cycle efficiencies of manures. Weather and nutrient inflows are the inputs for FARMSIM which are taken care of in simulations and altered for scenario analysis. Data derived from experimentation and standardized process-based models were employed to derive relationships functional in nature which are constructed to develop the submodels of Farming System Simulation Model (FARMSIM). Submodels introduce processes and interactions in a subjective manner and operate taking various time steps such as monthly for LIVSIM, HEAPSIM, and seasonal ones for FIELD.[82]

8.15 PRESENT SCENARIOS

A farming community of northeast Zimbabwe was selected, having 66 households, categorized into four resource groups (RGs) such as the livestock owners (RG1 and RG2) and the farmers with no livestock (RG3 and RG4); groups 1, 2 and 3, 4 have different land and labor available to them. Farmers differentiate fields in three types, namely, homefields, midfields, and outfields, which are different in terms of fertility and inputs amounts they receive. Homefields are the fields around the house, mid and outfields are infertile fields a few hundred meters to a few kilometers away from the house.

Two types of scenarios were assessed in time:

- a baseline scenario in which actual farmers practices were assessed, and
- target fertilized scenario, where inorganic fertilizer and organic resources are increasingly used uniformly across field types, and manure application in outfields is increased too.

Fertilizer rates of 60 kg N ha^{-1} and 30 kg P ha^{-1} employed during the simulations were the optimized rates during the experimentation.[83] Home-fields of RG1 and RG2 got half of the fertilizer to simulate fertilizers for maintenance. Comparatively in RG3 and RG4, crop residues were applied in the soil and fertilizers added across all fields. Crop residues from mid and outfields of all RG were considered to be grazed by livestock freely roaming. Scenarios are assessed based on the food grain and livestock production, agronomical N efficiency, and N budgets that are partial.

8.16 FINDINGS

Increased use of fertilizer and organic resources fed at lower rates in poor fields have increased grain production in the outfields of rich farmers (RG1 and RG2) and in home and outfields of poor farmers (RG3 and RG4). In the homefields of rich farmers, producing almost yield obtainable, the targeting fertilizers have increased agronomic efficiency and lowered the N balance partially as against the baseline. N agronomical efficiencies were increased for all fields, and there was an upward trend over time for the target fertilizers. Highest rises in agronomic efficiencies were recorded in the fields that respond, like the outfields of rich and the home fields of poor farmers. N balances altered a bit or reduced marginally for all fields and resource groups.

The responses to fertilizers are in tandem with the changes in soil carbon, derived from the organic resources management. Though soil C contents reduced in almost all soils in the baseline due to cultivation continuously with small inputs, C contents of homefields of rich farmers raised due to the increased levels of manure, fertilizers, and crop residues. In the target fertilizer scenario, C contents of all fields of rich farmers enhanced, and the homefields of poor farmers. The rate of degradation on long-term basis of outfields of poor farmers is lowered down in the target fertilizer scenario. Little increase in soil C contents because of investment in C management and nutrient flows increased land production at the farm level and the food self-sufficiency ratio, producing grain surpluses for the poor farmers for marketing. However, the effect of fertilizer in improving land productivity is largely related to the use of common grazing lands for feeding livestock to generate manure. A little proportion of manure excreted N (i.e., 14–16%) is subjected to on farm recycling and utilized for crop production. The remainder is partially lost from grazing lands, on farm from stalls and in manure handling. N budgets partially at the farm level are positive for rich farmers (RG1 and RG2) and abysmally low for poor farmers (RG3 and RG4). Increase in the use of fertilizers raises production but may rise the extent of N lost from farm with the produced food grain. This asks for assessment approaches for enhancing food production from the economical and environmental perspectives, such that farmers utilize the costlier fertilizers with efficiency without polluting their environment. Other studies have reported that N recycling in CLIs shows that a small degree (<10.5%) of all N flows.[84] Increasing cycling efficiency in a marginal holder system is not easy to achieve, largely because of labor shortages and priority of farmers in short-term returns for their investments.

8.17 USE OF ENHANCED CROP–LIVESTOCK INTEGRATION FOR CHECKING NITROGEN SURPLUS IN DAIRY INDUSTRY

Due to separation from continents and the arable land available to limited extent, agricultural development of La Réunion Island is facilitated by financial and technical assistance from the European Union. Since earlier part of 2000s, subsidies on feeds and mineral fertilizers, promoted the rapid intensification of dairy but did not give a boost to recycling of on-farm byproducts like organic fertilizers and forages. In 2004, farms had an average productivity of 5740 kg of milk cow^{-1} year^{-1}. Four thousand cows moved over 136 farms produced 23 m liters of milk. The local milk production covered around 29% of the consumption of the island (845,000 residents), the remainder was covered by imports in the form of milk powdered and cheese. This kind of intensification relying upon high on farm storing rates (average 4.5 livestock ha^{-1}) and on high input consumption (average 11.5 kg of feeds cow day^{-1}) produces larger large farm gate nutrient excess (average 478 kg N ha^{-1} year^{-1}) that increases environmental problems.[85] The farm model Global Activity Model for Evaluating Dairy Enterprises (GAMEDE) was designed, along with farmers, alternative management approaches that could lower down environmental risks related to N surplus because of farming. Crop livestock integration (CLI) at the farm level is assumed to be a strategy-holding lot much of promise.

8.18   METHODOLOGY USED

The organic and mineral fertilizer management module, GAMEDE is a simulation model which is dynamic and represents the N cycle of an agro-ecosystem in a farm. The model is instrumental in integrating decision-making and biophysical processes simulated representing interactions on a daily time basis. The major system parts are the farmer, forage crops, soils, and livestock. Major biophysical modules are dependent upon current mechanistic models such as monocalcium phosphate (MCP), Mosaics Cropping System (MOSAICS), In Ration software (INRATION), Cornell Net Carbohydrate and Protein *System* (CNCPS), a Decision Support System for the Management of Rotational Grazing in a Dairy Production (SEPATOU), and GRAZEIN.[86] GAMEDE simulates daily N gaseous losses and the variations in the N contents from fertilizers during their management in the steps like manure collection in barns, storage, composting, and application.

Four different manure management systems (MMS) in use are as follows:

(1) direct deposition when grazing is underway,
(2) manure in liquid form,
(3) raw manure in solid form, and
(4) solid manure in compost form.

In reality, MMS may coexist in a farm as a result of farmers' management of the livestock groups' excreta in a different way. Further, the same group of animals may add daily to the manure flow of various MMS. The total excreta produced by each animal group is divided between MMS in line with the time livestock spent during grazing or feeding in barns, as well as bedding practices. GAMEDE provides for each MMS which is considered as a sequence stage of handling in the form of chain of flows and stocks of different length: the shorter chain represents direct deposition during grazing and the longer one as composted solid manure. Irrespective of the MMS, it is crude effluent that is generated followed by either the application on crops or to soils. On the handling front, the N and dry matter of the manure are upgraded on daily basis in accordance with the feeding and bedding practices. Every fertilizer-handling step corresponds to an N emission flow computed based on the emission rates. The emission rates are determined from temperate climates data, while climatic correction indices are used to account for the La Réunion's tropical conditions and climatic variability.[87]

The indicator computed using GAMEDE to evaluate CLI in East African farm agro-ecosystems of the network analysis used to quantitate relationships within ecosystems and employed.[88] As highly integrated systems have more internal recycling and independent of the external environment, the extent of crop–livestock integration (DCLI) is computed as follows:

$$DCLI = 1 - \frac{N_{in}}{TST}$$

where N_{in} (kg N year^{-1}) is the amount of N that is made available from the external environment into the system and The Total System (TST) the flow through total system (kg N year^{-1}), or the sum of all N flows in the agro-ecosystem.

8.19   PRESENT SCENARIOS

GAMEDE was applied for the simulation of six farms, representing the management practices diversity and the type of soil and climate conditions prevailing on the island. Further simulations[89] involve four scenarios for differentiation. The "baseline scenario" shows present management practices of the farm. Ensiling of grass is the single method used on the farm of forage harvest. Livestock excreta are in particular taken care of in the form of slurry, containing 70% excreted indoors.

Grasslands are fed mainly with more quantities of mineral fertilizers such as one containing 264 kg of mineral and 55 kg of manure N year^{-1} ha^{-1} of cut grassland. On-farm generated organic fertilizers are underutilized to a large extent. The risk of overflow of slurry pit is large, in particular during the rainy season. The extent to which feeds fed to cows is low (<10 kg flowering maturity (FM) from concentrate cow day^{-1}) as against dairy systems in La Réunion. Feed self-sufficiency is therefore high (approx. 46%). In the "raised recycling of manure" as scenario 1, lactating cows are provided with cubicles on the advice of dairy industry experts. Cows stay in the barns throughout day and exercise regions become grasslands for grass silage leading to collection of manure further efficiently. As farm 3 is located in the rainfed area of the island, the pit containing slurry is covered to contain overflows and dilution.

Larger quantities of concentrated slurry are made available to fertilize grassland. The slurry is necessarily spread on all the farm's grasslands (not presently practiced) to raise the quantity of slurry employed to check overflows. The application rates of slurry are raised to homogenize fertilizer level between various farm plots. More manure and concentrated too on grasslands present farmer 3 the opportunity not to use mineral fertilizers without cutting down forage production. In scenario 2, the excess on farm produced grass silage leading to raised recycling of manure in scenario 1 is employed to feed animals and to lower down the amount of imported sugar cane straw as a feed. The silage surplus, obtained by substituting sugar cane straw with grass silage is available for heifers and cows. Further, farmer 3 presently is not adaptable to the quantity of concentrate feeds fed to lactating cows in accordance with their stage of lactation. The "improved herd reproductive performance" scenario 3, calving interval reduces to 380 days which can be simulated, judged realistically by local experts. Keeping cows in barns facilitates heat detection, as the farmers spend extra time looking after their cows. As against present herd performance, this would lead to a lowering down the calving interval by 18% for farm 3.

8.20 SIMULATION FINDINGS

The liberation of N gaseous emissions and losses because of overflows from slurry pit over 3 years for 0 and 1 scenarios. The effect of 0–3 scenarios on farm self-sufficiency and its extent of CL_i, on N flows, animal and crop yield. Increase in recycling of organic fertilizers generated on-farm (scenario 1), enhances fertilizer self-sufficiency (29.6%) and

N surplus (−22.5%). The increase in the use of slurry makes a larger use of nutrients and organic materials to soils and furthers the farm forage productivity by 3.8%. Sound slurry management reduces N losses with the help of overflows by 96%, but raises N gaseous emissions by 24.5%, particularly during use. Enhanced recycling of forages generated on-farm (scenario 2) raises farm feed self-sufficiency (+9.65%) and thereby adds up to N surplus furtherance (−9.9%), as more production of forage brings about lowering down of imports of conserved forages. Application of the concentrate feeds fed to lactating cows in accordance with their lactation stage enhances the efficiency in the utilization of concentrate (+13.2%). Increase in herd reproductive performance cuts down the time wherein cows get unproductive, and consequently milk productivity is increased by 8.5%. Improved recycling of organic fertilizers and forages generated on-farm, in combination with the rise in herd reproductive performance, further the farm self-sufficiency and its extent of CLI (+16.5%). This results in a reduction in the N leaching risk. Slurry pit overflows and subsequently runoff are lowered down. The farm N use efficiency is ultimately increased by 38.5%, leading to an N surplus lower than 350 kg N ha^{-1} year^{-1}.

8.21 GLOBAL VIEW OF NITROGEN LOSSES AND WORLD AGRO-SYSTEMS

As a result of the key role of food production in the modification of the N cycle at global level, a host of studies have generated views of the terrestrial N cycle with agriculture at the center stage[90] taking into account the global N flows with respect to crop production for feed and food. A similar reasoning for animal production, along with pastoral systems, was applicable. Such studies were made use of in detail by FAO[91] to assess effect on the environment of the livestock on global front. In the recent past, a higher resolution evaluation of global N flows in cropland was undertaken. The figures of such studies to take a view of the global agricultural front in accordance with the concept-based flows stock model employed to showcase agro-ecosystems in both developed[91] and developing world. The simplified global N cycle for the agricultural sector on the whole for the mid-1990s excluding forestry. It measures the major N losses in unregulated flows and its major N exchanges with other anthropological interventions (in particular provision of food and consumption of mineral fertilizer).

The global N losses from agroecosystems across the world total 7980 MT N year^{-1} which is equivalent to 40.1 MT NH_3 and 2.9 MT N_2O through gaseous emissions, 18 MT NO_3^- by leaching, and 20.1 MT organic N through runoff. It shows that 15.5% of applied N as mineral fertilizers and 30.5% of N excreted via pet animals globally is lost in terms of harmful NH_3 and N_2O and unavailable to plants used by animals and humans as feed. N losses are strongly connected with the availability of livestock metabolic activities in the cycle. Eighty-five percent of total agricultural N losses, 72% of NH_3, and 73% of N_2O are concerned with manure management. NH_3 losses at global level were determined from agriculture to be, housing and storage combined and pastoral systems 12–27%, from grazing 6% to 16%, from manure spreading in cropland and grassland 10–30%, and from N fertilizer use in cropland 10–19%. Forty-four percent of N available on earth (as mineral fertilizers and excreted by pets) is lost before taken up by plants. Thus, there is ample scope for growth from the conservation of N at global level agro-ecosystems perspective.

There are innumerable technical initiatives for compensating N losses to the environment both at system and process levels. There are no studies regarding evaluation of what would be the potential headway made by combining all such technologies worldwide. Such an assessment has become complex due to the diversity of agricultural systems and the types of soils, climates, and socio-economical situations confronted by the world. A better option would be to utilize typologies suggested by FAO regarding farming systems worldwide to make out representative areas and systems to be studied on priority. The analysis of such cases would be extended to assess what could be the worldwide gains from food production and environmental perspective in association with the enhanced N management in agro-ecosystems. Details exist to different levels in the typologies proposed, such as 11 livestock and 72 farming systems compared worldwide[92] for the developing countries. The degree of detail is defined from data available on nutrient losses and their potential loss at process (animal–manure–soil–plant) and system (farm) levels. Although liberation rates and factors controlling N losses are comparatively documented in the sound manner for temperate countries, standards for tropical contexts are unavailable. Further, the main technologies are designed for temperate and intensive systems and are not adaptable for low input systems came across by tropical countries. Majority of farmers do not have access to such technologies at this juncture.

Tremendous progress can be made by varying farming practices integrated with farm gate nutrient balances[93] and network analysis indicators to analyze practical systems. Integration with nutrient flow-stock models for simulating substitutional farming systems and exploring various coupled techniques and enhanced practices. In both tropical and temperate situations, a preferred use of organic fertilizers will boost global N use efficiency followed by rise in the yield of farming systems and reduce the potential risk to the environment. N dynamics in agro-ecosystems in temperate climates are comparatively documented in sound fashion however are not studied comprehensively in tropical nations. Simplification of experimental methods like mass-based ones may be of great help to develop a database rapidly and in cost-effective manner. Nutrient flow models are these days comparatively very common in developed nations[94–98] but are scanty in developing nations. Although size-dependent uncertainties on the flows are probably more in tropical countries, some simplified models to synthesize already available data and identify unavailable data from the point of view of evaluating the global contribution made by the enhanced nutrient management in the production of green food for the human population worldwide.

8.22   CONCLUSION

Farming agriculture with small landholdings in southern Africa should increase the amount of nutrient flows to better food self-sufficiency as well as security which may bring in environmental issues in the time to come, as long as agriculture does not generate more surpluses of N, as compared to developed world.[99] However, there is scope for exploring cost-effective options for the management of soil fertility and thereby contributing to improve food security.

In farm 3, simulation showed that variations in practices for increasing CLI and farm self-sufficiency lead to cut down the risk of nutrient leaching of La Réunion dairy farms even on increasing recycling of manure improves N gaseous losses simultaneously. This modeling shows the role of indirect mitigation options, livestock feeding, and reproduction. Furthermore, indicators in the complete analysis[100] also showcase the positive effect of raised N recycling on economic efficiency of farm.

As explained in this chapter for the nitrogen (N), the sources of nutrient losses within the nutrient cycle in agro-ecosystems are many and losses are large. For example, it is determined that 15.5% of used N as mineral fertilizers and 30.5% of excreted N by pets worldwide is lost as harmful gases (NH_3 and N_2O). Innumerable technical options are in existence to contain these losses and are related to animal husbandry and enhanced manure management techniques. But on applying in combination, options interact, and then efforts put in at one stage may be compensated but negatively in another stage (e.g., checked emissions during collection results into more emissions during storage and postapplication). As a result a comprehensive view of the nutrient cycling in the agro-ecosystems is required, with two farm-scale dynamic simulation models. Simulation findings predict that management of nutrient cycles may have a larger impact on enhancing land productivity and food production in low-input systems, and lowers environmental impact of high-input systems. Agricultural systems found worldwide are diverse and present models are not able to accommodate this diversity. Science contributes to produce simplified and applicable tools that help practitioners to assess nutrient use efficiencies of the systems. Tropical areas will be the field of major change in agriculture in the coming decades, and they are paradoxically the not well-documented areas on biophysical and managerial processes controlling nutrient losses. Research contributes to exploring nutrient cycles and opportunities to improve sustainable food production.

The contribution made by agricultural sector which is developing to global environmental transition is based preferably on projections, because biophysical characterization of nutrient cycles grounded on field observations and measurements is unavailable. This information is required to assess the global potential contribution of nutrient cycling and conservation in the production of sustainable food for the human population worldwide.

KEYWORDS

- nutrient cycle
- terrestrial ecosystem
- eutrophication
- stratospheric ozone layer

REFERENCES

1. Alvarez, S.; Salgado, P.; Vayssières, J.; Tittonell, P.; Bocquier, F.; Tillard, E. *Modelling Crop–livestock Integration Systems at a Farm Scale in a Highland Region of Madagascar: A Conceptual Model.* In Proceedings of SAPT International Conference, Guadeloupe, French West Indies, November 15–18, 2010.
2. Amon, B.; Amon, T.; Boxberger, J.; Alt, C. Emissions of NH_3, N_2O, and CH_4 from Dairy Cows Housed in a Farmyard Manure Tying Stall (Housing, Manure Storage, Manure Spreading). *Nutr. Cycl. Agroecosyst.* **2001,** *60,* 103–113.
3. Amon, B.; Amon, T.; Boxberger, J.; Pöllinger, A. *Emissions of NH_3, N_2O and CH_4 from a Tying Stall for Milking Cows, During Storage and Farmyard Manure and After Spreading.* In Proceedings of 8th International Conference on "Management Strategies for Organic Waste in Agriculture", Rennes, France, May 26–29, 1998.
4. Amon, B.; Moitzi, G.; Schimpl, M.; Kryvoruchko, V.; Wagner-Alt, C. *Methane, Nitrous Oxide and Ammonia Emissions from Management of Liquid Manures, Final Report 2002.* On behalf of "Federal Ministry of Agriculture, Forestry, Environmental and Water Management" and "Federal Ministry of Education, Science and Culture," 2002.
5. Andrieu, N.; Nogueira, D. M. Modeling Biomass Flows at the Farm Level: A Discussion Support Tool for Farmers. *Agron. Sustain. Dev.* **2010,** *30,* 505–513.
6. Baggs, E. M.; Philippot, L. Microbial Terrestrial Pathways to N_2O. In *Nitrous Oxide and Climate Change*; Smith., K. A., Ed.; Earthscan: Abingdon, 2010.
7. Beusen, A. H. W., Bouwman, A. F.; Heuberger, P. S. C.; Van Drecht, G.; Van Der Hoek, K. W. Bottom-up Uncertainty Estimates of Global Ammonia Emissions from Global Agricultural Production Systems. *Atmos. Environ.* **2008,** *42,* 6067–6077.
8. Børsting, C. F.; Kristensen, T.; Misciattelli, L.; Hvelplund, T.; Weisbjerg, M. R. Reducing Nitrogen Surplus from Dairy Farms: Effects of Feeding and Management. *Livest. Prod. Sci.* **2003,** *83,* 165–178.
9. Bouwman, A. F.; Boumans, L. J. M.. Emissions of N_2O and NO from Fertilised Fields: Summary of Available Measurement Data. *Global Biogeochem. Cycles* **2002a,** *16* (4), 6.
10. Bouwman, A. F.; Boumans, L. J. M. Modeling Global Annual N_2O and NO Emissions from Fertilised Soils. *Global Biogeochem. Cycles* **2002b,** *16* (4), 28.

11. Bouwman, A. F.; Beusen, A. H. W.; Billen, G. Human Alteration of the Global Nitrogen and Phosphorus Soil Balances for the Period 1970–2050. *Global Biogeochem. Cycles* **2009,** *23* GB0A04.

12. Bouwman, A. F.; Lee, D. S.; Asman, W. A. H.; Dentener, F. J.; Van Der Hoek, K. W.; Olivier, J. G. J. A Global High-Resolution Emission Inventory for Ammonia. *Global Biogeochem. Cycles* **1997,** *11* (4), 561–587.

13. Breman, H.; Groot, J. J. R.; Van Keulen, H. Resource Limitations in Sahelian Agriculture. *Global Biogeochem. Cycles* **2001,** *11,* 59–68.

14. Brouwer, J.; Powell, J. M. Increasing Nutrient Use Efficiency in West African Agriculture: The Impact of Micro-Topography on Nutrient Leaching from Cattle and Sheep Manure. *Agric., Ecosyst. Environ.* **1998,** *71,* 229–239.

15. Burton, C.; Jaouen, V.; Martinez, J. Traitement des effl uents d'élevage des petites et moyennes exploitations. *Guide technique à l'usage des concepteurs, bureaux d'études et exploitants*; QUAE: Versailles, 2007.

16. Bussink, D. W.; Oenema, J. Ammonia Volatilization from Dairy Farming Systems in Temperate Areas: A Review. *Nutr. Cycl. Agroecosyst.* **1998,** *51,* 19–33.

17. Carpenter, S. R.; Caraco, N. F.; Correll, D. L.; Howarth, R. W.; Sharpley, A. N.; Smith, V. H. Nonpoint Pollution of Surface Waters with Phosphorus and Nitrogen. *Ecol. Appl.* **1998,** *8,* 559–568.

18. Cassman, K. G.; Dobermann, A.; Walters, D. T. Agroecosystems, Nitrogen-Use Efficiency, and Nitrogen Management. *AMBIO* **2002,** *31* (2), 132–140.

19. Chadwick, D. R.; Matthews, R.; Nicholson, R. J.; Chambers, B. J.; Boyles, L. O. *Management Practices to Reduce Ammonia Emissions from Pig and Cattle Manure Stores*. In RAMIRAN 2002: Proceedings of 10th International Conference on "Recycling of Agricultural, Municipal and Industrial Residues in Agriculture," Strbskè Pleso, Slovakia, May 14–18, 2002.

20. Chardon, X.; Rigolot, C.; Baratte, C.; Le Gall, A.; Espagnol, S.; Martin-Clouaire, R.; Rellier, J.-P.; Raison, C.; Poupa, J.-C.; Faverdin, P. *Melodie: a Whole-Farm Model to Study the Dynamics of Nutrients in Integrated Dairy and Pig Farms*. In MODSIM 2007, Proceedings of the International Congress on Modelling and Simulation: "Land, Water and Environmental Management: Integrated Systems for Sustainability," Christchurch, New Zealand, December 10–13, 2007.

21. Chaumet, J. M.; Delpeuch, F.; Dorin, B.; Ghersi, G.; Hubert, B.; Le Cotty, T.; Paillard, S.; Petit, M.; Rastoin, J. L.; Ronzon, T.; Treyer, S. *Agrimonde: Agricultures et alimentations du monde en 2050: scénarios et défi s pour un développement durable*. CIRAD/INRA, 2009.

22. Chen, G. F.; Cao, L. Y.; Wang, G. W.; Wan, B. C.; Liu, D. Y.; Wang, S. S. Application of a Spatial Fuzzy Clustering Algorithm in Precision Fertilisation. *N. Z. J. Agric. Res.* **2007,** *50* (5), 1249–1254.

23. Chikowo, R.; Mapfumo, P.; Nyamugafata, P.; Giller, K. E. Mineral N Dynamics, Leaching and Nitrous Oxide Losses Under Maize Following Two Years Improved Fallows on a Sandy Loam Soil in Zimbabwe. *Plant Soil* **2004,** *259,* 315–330.

24. Cros, M. J.; Duru, M.; Garcia, F.; Martin-Clouaire, R. A Biophysical Dairy Farm Model to Evaluate Rotational Grazing Management Strategies. *Agronomy* **2003,** *23,* 105–122.

25. Clark, M. S.; Horwath, W. R.; Shennan, C.; Scow, K. M.; Lantni, W. T.; Ferris, H. Nitrogen, Weeds and Water as Yield-Limiting Factors in Conventional, Low-Input, and Organic Tomato Systems. *Agric., Ecosyst. Environ.* **1999**, *73*, 257–270.

26. Clemens, J.; Ahlgrimm, H. J. Greenhouse Gases from Animal Husbandry: Mitigation Options. *Nutr. Cycl. Agroecosyst.* **2001**, *60*, 287–300.

27. De Klein, C. A.; Eckard, M. R.; van der Weerden, T. J. N_2O Emissions from the Nitrogen Cycle in Livestock Agriculture: Estimation and Mitigation. In *Nitrous Oxide and Climate Change*; Smith, K. A., Ed.; Earthscan: Abingdon, 2010.

28. Delagarde, R.; Faverdin, P.; Baratte, C.; Peyraud, J. L. *Prévoir l'ingestion et la production des vaches laitières: GrazeIn, un modèle pour raisonner l'alimentation au pâturage*. In Proceedings of the 11th International Symposium Rencontres Recherches Ruminants, Paris, December 8–9, 2004.

29. Di, H. J.; Cameron, K. C. Nitrate Leaching in Temperate Agroecosystems: Sources, Factors and Mitigating Strategies. *Nutr. Cycling Agroecosyst.* **2002**, *64*, 237–256.

30. Di, H. J.; Cameron, K. Mitigation of Nitrous Oxide Emissions in Spray-Irrigated Grazed Grassland by Treating the Soil with Dicyandiamide, a Nitrification Inhibitor. *Soil Use Manage.* **2003**, *19*, 284–290.

31. Dixon, J.; Gulliver, A.; Gibbon, D. *Farming Systems and Poverty: Improving Farmers' Livelihoods in a Changing World*; FAO/World Bank: Rome, 2001.

32. Dollé, J. B.; Capdeville, J.; Martinez, J.; Peu, P. *Emissions d'ammoniac en bâtiments et au cours du stockage des déjections en élevage bovin*. Institut de l'Elevage: Paris, 2000.

33. Dollé, J. B.; Robin, P. *Emissions de gaz à effet de serre en bâtiment d'élevage bovin*. In AFPF 2006, Seminar on "Prairies, élevage, consommation d'énergie et gaz à effets de serres", Paris, March 27–28, 2006.

34. Edwards, C. A.; Grove, T. L.; Harwood, R. R.; Pierce Colfer, C. J. The Role of Agroecology and Integrated Farming Systems in Agricultural Sustainability. *Agric., Ecosyst. Environ.* **1993**, *46*, 99–121.

35. EMEP-CORINAIR. *Emission Inventory Guidebook*; European Environment Agency: Copenhagen, 2001.

36. FAO/IFA. *Global Estimates of Gaseous Emissions of NH_3, NO and N_2O from Agricultural Land*; FAO: Rome, 2001.

37. Fath, B. D.; Patten, B. C. Review of the Foundations of Network Environ Analysis. *Ecosystem* **1999**, *2*, 167–179.

38. Faverdin, P.; Delagarde, R.; Delaby, L.; Meschy, F. *Réactualisation des équations du livre rouge: alimentation des vaches laitières*; Quae: Versailles, 2007; pp 23–55.

39. Fox, D. G.; Tedeschi, L. O.; Tylutki, T. P.; Russell, J. B.; Van Amburgh, M. E.; Chase, L. E.; Pell, A. N.; Overton, T. R. The Cornell Net Carbohydrate and Protein System Model for Evaluating Herd Nutrition and Nutrient Excretion. *Anim. Feed Sci. Technol.* **2004**, *112*, 29–78.

40. Frank, B.; Persson, M.; Gustafsson, G. Feeding Dairy Cows for Decreased Ammonia Emission. *Livest. Prod. Sci.* **2002**, *76*, 171–179.

41. Gac, A.; Béline, F.; Bioteau, T. Flux de gaz à effet de serre (CH_4, N_2O) et d'ammoniac (NH3) liés à la gestion des déjections animales: Synthèse bibliographique et élaboration d'une base de données. [Greenhouse Gases and Ammonia Emissions from

Livestock Wastes Management: Bibliographic Review and Database Development: Technical Report]. CEMAGREF, Paris.

42. Galloway, J. N.; Aber, J. D.; Erisman, J. W.; Seitzinger, S. P.; Howarth, R. W.; Cowling, E. B.; Cosby, B. J. The N cascade. *BioScience* **2003**, *53*, 341–356.

43. Galloway, J. N.; Dentener, F. J.; Capone, D. G.; Boyer, E. W.; Howarth, R. W.; Seitzinger, S. P.; Asner, G. P.; Cleveland, C. C.; Green, P. A.; Holland, E. A.; Karl, D. M.; Michaels, A. F.; Porter, J. H.; Townsend, A. R.; Vöosmarty, C. J. Nitrogen Cycles: Past, Present, and Future. *Biogeochemistry* **2004**, *70*, 153–226.

44. Galloway, J. N.; Schlesinger, W. H.; Levy II, H.; Michaels, A.; Schnoor, J. L. Nitrogen Fixation: Anthropogenic Enhancement—Environmental Response. *Global Biogeochem. Cycles* **1995**, *9*, 235–252.

45. Gangbazo, G.; Pesant, A. R.; Cluis, D.; Couillard, D.; Barnett, G. M. Winter and Early Spring Losses of Nitrogen Following Late Fall Application of Hog Manure. *Can. Agric. Eng.* **1995**, *37*, 73–79.

46. Simon, J. C.; Grignani, C.; Jacquet, A.; Le Corre, L.; Pagès, J. **2000**. Typology of Nitrogen Balances on a Farm Scale: Research of Operating Indicators. *Agronomy Sustain. Develop.* 20, 175–195.

47. Giller, K. E. *Nitrogen Fixation in Tropical Cropping Systems*. CAB International: Wallingford, 2001.

48. Harper, L. A.; Sharpe, R. R.; Parkin, T. B. Gaseous Emissions from Anaerobic Swine Lagoons: Ammonia, Nitrous Oxide, and Dinitrogen Gas. *J. Environ. Qual.* **2000**, *29*, 1356–1365.

49. Hartung, J.; Phillips, V. R. Control of Gaseous Emissions from Livestock Buildings and Manure Stores. *J. Agric. Eng. Res.* **1994**, *57*, 173–189.

50. Hassouna, M.; Robin, P.; Brachet, A.; Paillat, J. M.; Dollé, J. B.; Faverdin, P. *Development and Validation of a Simplified Method to Quantify Gaseous Emissions from Cattle Buildings*. In Proceedings of the XVIIth World Congress of the International Commission of Agricultural and Biosystems Engineering (CIGR), Québec City, Canada, June 13–17, 2010.

51. Herrero, M.; Thornton, P. K.; Notenbaert, A. M.; Wood, S.; Msangi, S.; Freeman, H. A.; Bossio, D.; Dixon, J.; Peters, M.; van de Steeg, J.; Lynam, J.; Rao, P. P.; Macmillan, S.; Gerard, B.; McDermott, J.; Sere, C.; Rosegrant, M. Smart Investments in Sustainable Food Production: Revisiting Mixed Crop–Livestock Systems. *Science* **2010**, *327*, 822–825.

52. (a) Hill, M. J. *Nitrates and Nitrites in Food and Water*; Ellis Horwood: Chichester, 1991. (b) Hooda, P. S.; Edwards, A. C.; Anderson, H. A.; Miller, A. A Review of Water Quality Concerns in Livestock Farming Areas. *Sci. Total Environ.* **1991**, *250*, 143–167.

53. IPCC. *Revised 1996 IPCC Guidelines for National Greenhouse Gas Inventories— Reference manual* (Vol. 3), IPCC: Geneva, 1997. http://www.ipcc-nggip.iges.or.jp/public/gl/invs6.html.

54. IPCC. *2006 IPCC Guidelines for National Greenhouse Gas Inventories—Agriculture, Forestry and Other Land Use* (Vol. 4), IPCC: Geneva, 2006. http://www.ipcc-nggip.iges.or.jp/public/2006gl/vol4.html.

55. Jarrige, R. *Ruminant Nutrition: Recommended Allowances and Feed Tables*; INRA: Versailles, 1989.

56. Jongbloed, A. W.; Lenis, N. P.; Mroz, Z. Impact of Nutrition on Reduction of Environmental Pollution by Pigs: An Overview of Recent Research. *Vet. Q.* **1997**, *19*, 130–134.

57. Kamukondiwa, W.; Bergström, L. Nitrate Leaching in Field Lysimeters at an Agricultural Site in Zimbabwe. *Soil Use Manage.* **1994**, *10*, 118–124.

58. Karisson, S.; Salomon, E. *Deep Litter Manure to Spring Cereals—Manure Properties and Ammonia Emissions.* In RAMIRAN 2002: 10th International Conference on "Recycling of Agricultural, Municipal and Industrial Residues in Agriculture," Strbskè Pleso, Slovakia, May 14–18, 2002.

59. Lague, C.; Fonstad, T. A.; Marquis, A.; Lemay, S. P.; Godbout, S.; Joncas, R. *Greenhouse Gas Emissions from Swine Operations in Québec and Saskatchewan: Benchmark Assessments*; Climate Change Funding Initiative in Agriculture (CCFIA), Canadian Agricultural Research Council: Ottawa, 2004.

60. Lavelle, P.; Bignell, D.; Austen, M.; Giller, P.; Brown, G.; et al. Vulnerability of Ecosystem Services at Different Scales: Role of Biodiversity and Implications for Management. In *Sustaining Biodiversity and Functioning in Soils and Sediments*, Wall, D. H., Ed.; Island Press: Washington, DC, 2004.

61. Leontief, W. W. *The Structure of American Economy, 1919–1939: An Empirical Application of Equilibrium Analysis*; Oxford University Press: New York, 1951.

62. Leteinturier, B.; Oger, R.; Buffet, D. *Rapport Technique sur le Nouveau Module de Croissance Prairiale.* Centre de Recherches Agronomiques de Gembloux, Belgique, 2004; p 37.

63. Levasseur, P.; Boyard, C.; Vaudelet, J. C.; Rousseau, P. *Evolution de la valeur fertilisante du lisier de porcs au cours de la vidange de la fosse de stockage, influence du brassage.* In Proceedings of the 31th Journées Recherche Porcine en France, Institut Technique du Porc, INRA, February 2–4, 1999.

64. Liu, J.; You, L.; Amini, M.; Obersteiner, M.; Herrero, M.; Zehnder, A. J. B.; Yang, H. A High Resolution Assessment on Global Nitrogen Flows in Cropland. *PNAS* **2010**, *107* (17), 8035–8040.

65. Marini, J. C.; Van Amburgh, M. E. Nitrogen Metabolism and Recycling in Holstein Heifers. *J. Anim. Sci.* **2003**, *81*, 545–552.

66. Marquis, A. *Emissions de gaz à Effet de Serre par les Animaux aux Bâtiments.* In Proceedings of the 65th Congress de l'Ordre des Agronomes du Québec, 2002.

67. Moller, H. B.; Sommer, S. G.; Anderson, B. H. Nitrogen Mass Balance in Deep Litter During the Pig Fattening Cycle and During Composting. *J. Agric. Sci.* **2000**, *137*, 235–250.

68. Martiné, J. F. Modélisation de la Production Potentielle de la Canne à Sucre en Zone Tropicale, Sous Conditions Thermiques et Hydriques Contrastées: Application du Modèle. Ph.D. Thesis, INA-PG, Paris, 2003; 116 p.

69. Monteny, G. J.; Bannink, A.; Chadwick, D. Greenhouse Gas Abatement Strategies for Animal Husbandry. *Agric., Ecosyst. Environ.* **2006**, *112*, 163–170.

70. Morvan, T.; Leterme, P. Vers une prévision opérationnelle des flux d'azote résultant de l'épandage de lisier: Paramétrage d'un modèle dynamique de simulation des transformations de l'azote et des lisiers. *Ingénieries* **2001**, *26*, 17–26.

71. Mosier, A.; Wassmann, R.; Verchot, L.; King, J.; Palm, C. Methane and Nitrogen Oxide Fluxes in Tropical Agricultural Soils: Sources, Sinks and Mechanisms. *Environ., Dev. Sustainability* **2004,** *6,* 11–49.

72. Mosley, J. C.; Cook, P. S.; Griffis, A. J.; O'Laughlin, J. Guidelines for Managing Cattle Grazing in Riparian Areas to Protect Water Quality: Review of Research and Best Management Practices Policy. Report No. 15 Policy Analyses Group, College of Forestry, Wildlife, and Range Sciences, University of Idaho, Moscow, ID, 1997.

73. Nevens, F.; Verbruggen, I.; Reheul, D.; Hofman, G.. Farm Gate Nitrogen Surpluses and Nitrogen Use Efficiency of Specialized Dairy Farms in Flanders: Evolution and Future Goals. *Agric. Syst.* **2006,** *88,* 142–155.

74. Nicks, B.; Laitat, M.; Vandenheede, M.; Desiron, A.; Verhaege, C.; Canart, B. Emissions of Ammonia, Nitrous Oxide, Methane, Carbon Dioxide, and Water Vapor in the Raising of Weaned Pigs on Straw-Based and Sawdust-Based Deep Litters. *J. Appl. Anim. Res.* **2003,** *52,* 299–308.

75. Nyamangara, J.; Bergström, L. F.; Piha, M. I.; Giller, K. E. Fertilizer Use Efficiency and Nitrate Leaching in a Tropical Sandy Soil. *J. Environ. Q.* **2003,** *32,* 599–606.

76. Nyamangara, J.; Piha, M. I.; Kirchmann, H. Interactions of Aerobically Decomposed Manure and Nitrogen Fertiliser Applied to Soil. *Nutr. Cycl. Agroecosyst.* **1999,** *54,* 183–188.

77. Oenema, O.; Oudendag, D.; Velthof, G. L. Nutrient Losses from Manure Management in the European Union. *Livest. Sci.* **2007,** *112,* 261–272.

78. Oenema, O.; Witzke, H. P.; Klimont, Z.; Lesschen, J. P.; Velthof, G. L. Integrated Assessment of Promising Measures to Decrease Nitrogen Losses from Agriculture in EU-27. *Agric., Ecosyst. Environ.* **2009,** *133,* 280–288.

79. Paillat, J. M.; Robin, P.; Hassouna, M.; Leterme, P. Predicting Ammonia and Carbon Dioxide Emissions from Carbon and Nitrogen Biodegradability during Animal Waste Composting. *Atmos. Environ.* **2005,** *39,* 6833–6842.

80. Parkinson, R.; Gibbs, P.; Burchett, S.; Misselbrook, T. Effect of Turning Regime and Seasonal Weather Conditions on Nitrogen and Phosphorus Losses during Aerobic Composting of Cattle Manure. *Bioresour. Technol.* **2004,** *91,* 171–178.

81. Petersen, S. O.; Lind, A. M.; Sommer, S. G. Nitrogen and Organic Matter Losses During Storage of Cattle and Pig Manure. *J. Agric. Sci.* **1998,** *130,* 69–79.

82. Juhasz, A.; Leip, R.; Mihelic, T.; Misselbrook, J.; Martinez, F.; Nicholson, H.; Poulsen, D.; Provolo, G.; Sorensen, P.; Weiske, A. *Recycling of Manure and Organic Wastes—A Whole-Farm Perspective.* Proceedings (Vol. I) of the 12th RAMIRAN International Conference "Technology for Recycling of Manure and Organic Residues in a Whole-Farm Perspective." Aarhus, Denmark, 2006.

83. Powell, J. M.; Broderick, G. A.; Grabber, J. H.; Hymes-Fecht, U. C. Technical Note: Effects of Forage Protein-Binding Polyphenols on Chemistry of Dairy Excreta. *J. Dairy Sci.* **2010,** *92,* 1765–1769.

84. Predotova, M.; Bischoff, W.-A.; Buerkert, A. Mineral-Nitrogen and Phosphorus Leaching from Vegetable Gardens in Niamey. *Niger. J. Plant Nutr. Soil Sci.* **2010a,** *174* (1), 47–55.

85. Predotova, M.; Gebauer, J.; Diogo, R. V. C.; Schlecht, E.; Buerkert, A. Emissions of Ammonia, Nitrous Oxide and Carbon Dioxide from Urban Gardens in Niamey. *Niger. Field Crops Res.* **2010b,** *115,* 1–8.

86. Predotova, M. E.; Schlecht, A.; Buerkert, A. Nitrogen and Carbon Losses from Dung Storage in Urban Gardens of Niamey. *Niger. Nutr. Cycl. Agroecosyst.* **2010c,** *87,* 103–114.

87. Ramisch, J. J. Inequality, Agro-Pastoral Exchanges, and Soil Fertility Gradients in Southern Mali. *Agric., Ecosyst. Environ.* **2005,** *105,* 353–372.

88. Raun, W. R.; Johnson, G. V. Improving Nitrogen use Efficiency for Cereal Production. *Agron. J.* **1999,** *91,* 357–363.

89. Robin, P.; Hassouna, M.; Paillat, J. M. *Multi-Element Combined Methods (MECM) Increase the Reliability of Emission Factor Measurement.* In Proceedings of the International Workshop on Pork Production "Porcherie verte," Paris, May 25–27, 2005.

90. Rotz, C. A. Management to Reduce Nitrogen Losses in Animal Production. *J. Anim. Sci.* **2004,** *82,* 119–137.

91. Rotz, C. A.; Taube, F.; Russelle, M. P.; Oenema, J.; Sanderson, M. A.; Wachendorf, M. Whole Farm Perspectives of Nutrient Flows in Grassland Agriculture. *Crop Sci.* **2005,** *45* (6), 2139–2159.

92. Rufino, M. C.; Tittonell, P.; van Wijk, M. T.; Castellanos-Navarrete, A.; Delve, R. J.; de Ridder, N.; Giller, K. E. Manure as a Key Resource Within Smallholder Farming Systems: Analysing Farm-Scale Nutrient Cycling Efficiencies with the NUANCES Framework. *Livest. Sci.* **2007,** *112,* 273–287.

93. Rufino, M. C.; Herrero, M.; van Wijk, M. T.; Hemerik, L.; de Ridder, N.; Giller, K. E. Lifetime Productivity of Dairy Cows in Smallholder Farming Systems of the Highlands of Central Kenya. *Animal* **2009a,** *3,* 1044–1056.

94. Rufino, M. C.; Tittonell, P.; Reidsma, P.; López-Ridaura, S.; Hengsdijk, H.; Giller, K. E.; Verhagen, A. Network Analysis of N Flows and Food Self-sufficiency: A Comparative Study of Crop–Livestock Systems of the Highlands of East and Southern Africa. *Nutr. Cycl. Agroecosyst.* **2009b,** *85,* 169–186.

95. Rufino, M. C.; Dury, J.; Tittonell, P.; Van Wijk, M. T.; Herrero, M.; Zingore, S.; Mapfumo, P.; Giller, K. E. Competing Use of Organic Resources, Village-Level Interactions between Farm Types and Climate Variability in a Communal Area of NE Zimbabwe. *Agric. Syst.* **2010,** *104,* 175–190.

96. Sanchez, P. A. Soil Fertility and Hunger in Africa. *Science* **2002,** *295,* 2019–2020.

97. Schaber, J. *FARMSIM: A Dynamic Model for the Simulation of Yields, Nutrient Cycling and Resource Flows on Philippine Small-Scale Farming Systems*; Institut für Umweltsystemforschung, University of Osnabrück: Germany, 1997.

98. Schröder, J. J.; Jansen, A. G.; Hilhorst, G. J. Long-Term Nutrient Supply from Cattle Slurry. *Soil Use Manage.* **2005,** *21,* 196–204.

99. Scrimgeour, G. J.; Kendall, S. Consequences of Livestock Grazing on Water Quality and Benthic Algal Biomass in a Canadian Natural Grassland Plateau. *Environ. Manage.* **2002,** *29,* 824–844.

100. Seré, C.; Steinfeld, H.; Groenewold, J. World Livestock Systems: Current Status, Issues and Trends. Animal Production and Health: Paper No. 127, FAO: Rome, 1995.

CHAPTER 9

Bionanocomposite Materials and Their Applications

SHRIKAANT KULKARNI[*]

Department of Chemical Engineering, Vishwakarma Institute of Technology, 666, Upper Indira Nagar, Bibwewadi, Pune 411037, India

[*]Corresponding author. E-mail: shrikaant.kulkarni@vit.edu*

ABSTRACT

Bionanocomposite materials by virtue of their strength in terms of structural diversity hold lots of promise and potential in terms of widespread applications in diverse fields ranging from sensing to energy production. The structural diversity can be put to advantage ranging from carbon nanotubes to collagen. The diversity can further offer host of combinations of biomaterials derived from bionanocomposites. The compositional diversity is further of immense interest in designing materials with requisite shape, size, geometry, morphology to meet the specific challenges in biocatalysis on demand. However, structural diversity may lead to varied expectations often with lexicon and evolution of much of literature. However, these materials are yet to be explored to their full potential. It is attributed to the nanotoxicity regulatory constraints and disparity in performance in terms of specificity in biocatalytic activity. However, biocatalysts with well-defined architectures exposed to chemical environments in tune with their biological activity can help increase yield by way of enhancing the substrate or mediator diffusion. Further, the right kind of architecture may present a soundness in stability in physicochemical conditions which otherwise may stifle the performance of catalysts.

This chapter emphasizes conventional processing techniques in developing the bionanocomposites as the biocatalyst. Further, it explains innovative processing technologies like electrospinning or bioprinting to shape living matter which could, in our knowledge, hasten the spectrum of applications of bionanocomposite materials for biocatalysis.

9.1   BIONANOCOMPOSITES AS BIOCATALYSTS

Biocatalysis plays a vital role in bringing about host of processes both biological and industrial ones.[1] It has applications covering spectrum of areas like environmental remediation,[2] food industry,[3,4] pharmaceutical synthesis,[5] energy generation,[6,7] sensing,[8,9] and biocomputing.[10–12] The two key mechanisms evolved over the time involve the evolutionary screening which is naturally occurring and the directed evolution which has come to the fore recently are prevalent and further the cause of it is the molecular machinery structure[13–16] significant in the area of enzyme-driven catalysis.

Biocatalytic performance enhancement does not seem to be within the reach of the material scientists working in developing bionanocomposites but more of a task of a protein engineer. The rationale behind the elaboration of bionanocomposite materials devoted to biocatalysis lies in the need to enhance the activity, stability, and recyclability of biocatalysts responsible for key biochemical processes. In such a context, we can consider the development of bionanocomposite materials as a general strategy to transpose the bioreactor technology often based on reaction vessels where the physicochemical conditions required for the biocatalysis to occur are externally controlled into solid-state materials that are intrinsically compatible with the biocatalyst's activity. The materials that scientists approach aim at providing the biocatalytic moieties with a suitable chemical and/or biological environment that will facilitate the catalytic activity. The selected examples illustrate that a careful control over the biocatalysts' mechanical, chemical, and biological environment can match or in some cases even enhance the performances observed in more classical environments. The various approaches surveyed in this chapter focus on the use of bionanocomposite materials as host matrices where functional biocatalytic units enzymes, cells, and so on are arranged to maximize their performance. Much in the same way as in architecture, here, form follows function. Given the interdisciplinary nature of the subject of designing bionanocomposites for biocatalysis, we believe there are as many viewpoints as there are relevant scientific disciplines to the field of biocatalysis.

Here, we present a materials science perspective on the subject. To explore the intersection between the design of bionanocomposites and their application in the domain of biocatalysis, this chapter is structured in two main sections: form and function in bionanocomposite biocatalysts, and applications of bionanocomposite materials in biocatalysis.

The nature of the biocatalytic bionanocomposite materials implies the presence of different molecular entities most often macromolecules of biological origin; some degree of clarification is needed to identify which systems have been contemplated in this chapter and which ones were considered to lie out of its scope. One of the most common definitions of bionanocomposites relies on the presence of a naturally occurring polymer usually the continuous phase in combination with one, or more, reinforcing phases displaying at least one dimension in the nanometer scale.[17,18] Most bionanocomposites either natural or synthetic such as bone, mother of pearl, exfoliated-clay-reinforced biopolymers, or carbon nanotube (CNT)-reinforced biopolymers fall within this simple definition. The case of bionanocomposites for biocatalytic application, however, poses a particular challenge to such a definition. In fact, bionanocomposites used in biocatalytic applications are often composed of multiple biological macromolecular entities playing remarkably different roles. In general, throughout this chapter, we shall consider the bionanocomposite materials devoted to biocatalysis as the combination of a structural part and a functional part. The bionanocomposite or structural part is, as previously defined, composed of one or more reinforcing phases dispersed within a macromolecular continuous phase. The biocatalytic or functional part consists of a biological entity such as living cells, enzymes, or other catalytic biomolecules housed by the bionanocomposite host. Some exceptions to this architecture will be considered for the cases where the originality of the approach or the merit of the results justifies the detour. While the discussion regarding the nature of the two main building blocks is essential, the complex questions of how to process them into materials while preserving the biocatalytic activity cannot be overlooked. Two requirements will be considered for such a discussion. The first regards the technical constraints imposed by the biocatalyst's sensitivity to environmental conditions such as pH, temperature, solvent nature, and osmolarity. The second relates to the processing requirements such as agitation, temperature, and addition of surfactants needed to attain a successful dispersion of the nanofiller phase throughout the continuous matrix. These two major considerations should provide the reader with a glimpse of the technical obstacles to overcome when addressing the elaboration of the final materials. The second main section focuses on the applications of bionanocomposite materials in biocatalysis. The reader will be guided throughout some of the more relevant examples where bionanocomposites have been used to enhance activity of biocatalysts in applications as diverse as environmental remediation, energy, biosynthesis, or sensing.

9.2   ROLE OF BIOPOLYMERS IN BIONANOCOMPOSITES

Biopolymers are very important components in living organisms. They provide, among other functions, the mechanical stability for the vast majority of different cells and tissues. In the case of animal cells, the mechanical stability is provided by the extracellular matrix (ECM), a complex biological material mostly composed of proteins and/or polysaccharides, providing a support medium for cells. Such ECM presents diverse compositions, ranging from support proteins such as collagen, elastin, or fibroin to polysaccharides such as chitin. On the contrary, other eukaryotes such as fungi, algae, and higher plants display a markedly different architecture to ensure their mechanical stability. In these organisms, the cell is surrounded by a structuring layer of polysaccharides called the cell wall.[19] In higher plants, cellulose—a crystalline polysaccharide composed of glucose units linked through a $1 \rightarrow 4$ glycosidic bond—is the main component of the cell wall. Its elevated Young's modulus plays a critical role in the mechanical support needed for trees to attain their dimensions. Bacteria, regardless of the structural differences between Gram-positive and Gram-negative in terms of outer architecture, also present a cell wall. Their composition is mainly based on peptidoglycans (i.e., polysaccharide chains cross-linked by small peptides). In any of the preceding cases, not only do biopolymers play a fundamental structural role, but also they stand out as the most extended interface between living cells and their surrounding environment. Naturally, the role of structural biopolymers in biological matter is not limited to the mechanics of tissues. Especially in multicellular organisms, the ECM is known for its relevant role in morphogenesis, aging, and disease.[19,20] The unique combination of a structural role with the capacity to host cells or other active biological entities makes biopolymers a rational choice for the design of materials expected to directly interact with biocatalysts. In this sense, it is expected that a judicious choice of biopolymer may positively contribute to create a suitable biochemical environment around biocatalysts. In this section, we will briefly overview some selected biopolymers that have proven to be particularly suitable to build a biocompatible environment to host either living cells or active biocatalysts.

Since the groundbreaking work by Lim and Sun,[21] where alginate beads were designed to host pancreatic islet cells, polysaccharide found in seaweed has been thoroughly explored to host living mammalian cells,[22,23] yeast cells,[24] or enzymes.[25-28] Its popularity as an encapsulation agent is no doubt related to its cytocompatibility and is to the fact that it is easily available. However, part of the success of alginate in the domain can also be attributed to its ability to be easily cross-linked by divalent cations. In fact, the "egg in a box" conformation adopted by the polysaccharide chains around divalent cations has provided researchers with an instrumental tool to control alginate's solubility, mechanical properties, and ultimately to generate an interface separating the encapsulated species from the bulk. Lim and Sun's work is a notorious example where such ability is explored to encapsulate cells. The pancreatic cell islets encapsulated in the alginate beads were physically separated by a calcium cross-linked "membrane" from the surrounding environment. The formation of an outer cross-linked shell prevented the recognition and elimination of the encapsulated biological material by the patient's reticuloendothelial system. Meanwhile, the soluble alginate liquid core acted as a suitable medium for cell suspension and viability. Such an early example of the use of alginate for the microencapsulation of potentially biocatalytic species has laid ground for a wide range of biomaterials for microencapsulation.[29-31] Recently, Christoph et al. developed alginate-based solid-state materials for yeast encapsulation.

Their approach was based on ice templating of living *Saccharomyces cerevisiae* cells suspended in an alginate sol by unidirectional freezing followed by freeze–drying. The resulting solid-state macroporous foam was able to extensively keep yeast cell's viability despite the lyophilization treatment. In this case, alginate acts not only as a scaffold into which the yeast cells are housed but also as a cryoprotectant preventing major cellular damage during the freezing and low-pressure steps. Besides alginate, other polysaccharides have been largely used to shape the interface between bioactive and structural moieties. Rocha et al.[32] have encapsulated both transforming growth factor β1 and adipose-derived stem cells in carrageenan-based hydrogels for cartilage tissue engineering. Once again the gelling properties of the polysaccharide proved to be a critical parameter for its application as an encapsulation medium. Similar to alginate or carrageenan, other polysaccharides have been thoroughly used to develop biocompatible constructs for the encapsulation of biologically active entities. In particular, gellan gum has been used to create polysaccharide beads able to house gasoline-degrading bacteria for environmental applications.[33,34] In parallel, the same polysaccharide has also found successful application in the clinical field as cell delivery vectors.[35] Similarly, agarose[36] or hyaluronic acid[37] has been widely used for clinical applications where the delivery of key cellular entities was primordial. In some cases, the assembly of oppositely charged biomacromolecules such as chitosan and hyaluronic acid to generate polyelectrolyte complexes provides the most interesting environment for the encapsulation and delivery of biologically active components, as reviewed by Rinaudo.[38]

Proteins, though less common as encapsulation medium than polysaccharides, have also been used as host matrices for living cells and other biocatalytic entities. The following examples illustrate that the use of proteins has been mostly focused either in the biomedical field or in relation to the food processing industry. In fact, the use of proteins such as collagen, fibrin, or albumin for drug-delivery applications has evolved into the delivery of enzymes, or, in some cases, metabolically active cells.[39-42] In parallel to these medical applications, the food industry has devoted considerable effort to creating protein-based systems capable of housing and delivering biocatalytic species.[43-45] Regardless of the envisioned application, the nature of the encapsulated species, or the processing conditions, the chemical environment provided by biopolymers notably by those displaying gelation has long been considered as the most convenient choice for the maintenance of the biological activity of biocatalysts or other encapsulated biological species. Recent results have also shown that beyond the chemical specificity of biopolymers, the mechanics of the host matrix are also determinant in the morphology and activity of cell cultures.[46] Palchesko et al. have demonstrated the dependence between the mechanics of tissues and the cellular response by varying independently the nature and the mechanics of the ECM biopolymers. They have conducted experiments in corneal epithelium, a cell monolayer located in the posterior surface of the cornea cultured in several biopolymer components of the ECM such as laminin, collagen I, collagen IV, fibronectin, and combination of laminin and collagen IV. According to their results, the cells cultured on the collagen IV with an elasticity modulus of 50 kPa resembled the most to the biological tissue, displaying polygonal morphology as in native tissue. Other examples, such as the viability and metabolic activity of fibroblasts in collagen matrices, have also been reported. According to the authors, high-collagen concentrations translated into enhanced mechanical properties that promoted cell migration and proliferation.[47] To some extent, the previous examples illustrate the need for specific mechanical properties while keeping the chemical nature of the biopolymer fraction unchanged. One of the most successful strategies to fine tune the mechanics of biopolymers has been addressed by the inclusion of an inorganic often more rigid moiety into the matrix. This approach, heavily inspired by biological nanocomposite materials such as bone, nacre, or arthropod exoskeletons, has found a remarkable success in a wide variety of applications and use a large array of structural components. The following section describes some of the most widely used components for the reinforcement of bionanocomposite materials.

The inorganic fraction, in the case of biopolymers, is used as fillers in nanocomposites, has been investigated as interfaces with biological species. A wide variety of materials such as clay minerals, layered double hydroxides (LDHs), carbonates, metal oxides, or carbonaceous nanomaterials have been explored to build nanocomposite materials in the presence of biopolymers, with different impact on the mechanical, transport, or swelling properties, among others.[17,18] However, their relevance in building extended interfaces with functional biological components is far more restricted. In this section, we will discuss some of the key features surrounding the most commonly used nanomaterials in bionanocomposites for biocatalysis. One of the central aspects surrounding the applicability of the selected nanomaterials for the development of bionanocomposites for biocatalysis is their "biocompatibility." Here, the term biocompatibility is considered in a broad sense, meaning that the contact between the biocatalytic fraction and the filler will not induce the denaturation or inactivation of the biocatalyst. As seen earlier, the ability of the nanofiller to modify the mechanical properties of the composite material is critical to modify the cellular response to the environment. In addition, the ability to functionalize the said nanofillers allows for a surface chemistry versatility that widens the range of applications and can be used to minimize detrimental impact on the biological moiety. The synthesis conditions required to obtain the solids are also critical to determine how the association strategy between the reinforcing moiety and the biological entities will be conducted. Silica, for instance, is easily obtained by a sol–gel route under eminently soft conditions, which has opened the possibility to the integration of sensitive biological materials during the synthesis step. On the contrary, particulate materials like filler are commonly subjected to harsher synthetic pathways, thus preventing direct incorporation of bioactive species. This is the case for materials such as CNTs, whose synthesis conditions are incompatible with the integrity of most molecules of biological origin. Finally, the fillers' intrinsic properties such as their electronic conductivity are critical depending on the considered applications.

Clay minerals are discussed as one of the first relevant inorganic nano-materials to possess the conceivable catalytic action which plays a decisive role in abiogenesis.[48,49] Their ordered structure with better capacity of adsorption, concentrate organic substances, and to work as templates for polymerization are major advantages that put clays as future supports for abiogenesis.[50] Despite the earlier connections between clay minerals and biological material, their role in constructing bionanocomposites was not well pursued.[51–53] Pioneering work by Fukushima on polymer clay nano-composites has led the adaptation of the similar approach to mechanically dispersed clay minerals as reinforcing agents in biopolymer matrices.[54] Clay minerals can be classified roughly based on their morphology as layered and microfibrous clays. Delamination of the mineral tactoids in layered clays sometimes is obtained by simple cation-exchanging process. The interlayer cations are instrumental in the structural integrity of clay minerals can be substituted by molecules with equivalent charge. Such an ability to intercalate other molecular moieties in the space available results into the organoclays and later fully exfoliated nanocomposites formation depending upon individual clay layers.[55] The intercalation biopolymers with positive charges in the space between the layers of layered clays were not suggested until the intercalation of chitosan in montmorillonite had not been discussed. This approach provided for the production of a 3D network of chitosan and discrete clay sheets as well as truncating the inter-layer charge by regulating the quantity of intercalated biopolymer.[56] The design of potentiometric sensors that respond to the either anions or cations based on the quantity of intercalated chitosan by exploring innovative results.[57] Microfibrous clays like sepiolite and palygorskite cannot either be exfoliated or intercalated because of the inversions of silica tetrahedra in a regular order which ensures interribbon covalent bond. Furthermore, these clays do not offer negative charge made up by the interlayer cations in excess. The chemical tailoring possibilities are constrained as against layered clays, substantial effort has been taken such that these inorganic natural nanomaterials are compatible with various biopolymers.[58–64] To facilitate the better interaction of fibrous clays with biopolymers, the high silanol surface density of these materials responsible for their hydrophilicity is taken into account. Moreover, the presence of the silanol groups extensively on the outer-surface of the fibers provides for the covalent functionalization by grafting of alkoxysilane side chain. Surface modifications of such kind help tailor the chemical affinity between the biopolymer species and the clay particles.[65] However, naturally, the use of biomolecules and clays is not limited to preparation of composites. For example, the interaction between enzymes and clays has been explored a lot for host of applications in biocatalysis.[66] Another critical issue is related to the clay particles' stability in aqueous environment as the colloidal nature of clay particles favors their processing by solution/suspension techniques. It is quite convenient in the perspective of biomaterials processing because of improved stability of biomolecules in aqueous media. As a result, majority clay-based bionanocomposite materials synthesized from clay minerals and biomolecules are processed by aqueous-based techniques like micro-encapsulation, gelation, solvent casting, and so on.

LDHs have been widely used as the filler in the form of inorganic moiety in the development of bionanocomposite materials. Their layered structure has strong semblance with clay minerals; although the sheets differ in their composition, depending upon metal hydroxides than silicates. The main advantage of such type of materials is intercalation of different molecules between the various layers. But the peculiarity of LDH is that the layers acquire a positive charge that allows entrapment of anions. Such marked difference allows a varied kind of biomolecules and biopolymers, which possess negative charge, and would therefore have difficulty in associating with clay minerals. However, this is not the only comparative advantage of LDH. They are simple and cheap in synthesizing,[67] particularly by coprecipitation methods, and have also proven biocompatibility.[68] The limitation could be the sensitivity of the LDH in acidic conditions. However, this has been exploited to advantage when LDH is used as drug carriers, as it releases its content on arrival in the stomach. The interest lies in applications involving long-term encapsulation of biocatalytic species. One of the possible solutions is to use LDH coupled with a right polymer, in preference a biopolymer. The potential applications of LDH-based bionanocomposites are innumerable, for example, silica-based materials. However, there is not much of work regarding the combination of LDH biopolymers and the encapsulated biocatalysts for applications in frontier areas like biosensing, bioremediation, or chemical catalysis. The application of LDH in bionanocomposite is very well known.[69] Various anionic biopolymers right from alginic acid to xanthan gum in [Zn_2Al] LDH are combined by using coprecipitation method. Further, it was observed that there are many uses of functional biopolymers coupled with LDH,[70] exhibiting the outstanding versatility of these materials as a part of bionanocomposites. LDHs have also been widely employed to entrap enzymes, typically due to their anion exchanging capacities.[71] LDH in biocatalytic applications can be used quite extensively used for encapsulation of bioactive molecules and drug delivery[72] Moreover, LDHs have been coupled with different biopolymers for ensuring efficient drugs' encapsulation. For example, alginate and zein in a magnesium–aluminum LDH were employed to generate an ibuprofen carrier.[73] Similarly, a pectin–chitosan/LDH bionanohybrid was configured for encapsulating antiinflammatory drug targeting colon.[74] Such devices are adaptable to host drugs, biomolecules, and even whole cells to broaden the horizons of potential application areas.

9.3 SILICA

Silica is a central material for designing composite materials. Silica-based materials like colloidal silica have been extensively employed for the production of polymer-based nanocomposites beyond glass fibers as technically dispersed in biopolymers.[75-77] As in the case of clay minerals, silica has got lots of functionalization potential because of the existence of silanol groups at its surface which provides for modifying surface chemistry of the particles in accordance with the polymer nature using grafting of organosilane molecules. Few of the examples of such an approach were oriented toward the configuration of conductive polymer-based nanocomposites[77] and further extended for the development of nanocomposites devoted for the cause of improving the mechanical properties of polymer.[75] Before using colloidal particles of silica for designing nanocomposite materials, the advent of sol–gel chemistry had already paved the way for researchers to investigate the compatibility between silica and biomolecules. In fact, in 1971,[78] encapsulation of trypsin in a silica gel was reported. Further, a number of cases of biomolecules (or cells) encapsulation in sol–gel silica matrices were reviewed by some researchers.[79] For the encapsulation of living or biologic entities, lower temperature and milder process parameters are favored most. The two most popular and biocompatible pathways are the application of sodium silicates[80] and alkoxide, for example, tetraethoxysilane.[81]

The coupling with a biopolymer for ensuring highest biocompatibility was also incorporated too early. For example, in 1988,[82] both alginate and colloidal silica were used for immobilizing quite a broad range of biological entities namely, γ-globulin, bovine serum albumin, β-galactosidase as well as whole yeast cells (*S. cerevisiae*). Earlier work was devoted for the cause of studying the difference in performance, preservation or improvement between free and entrapped functionalized entities, but not for exploring the potential applications.[83–86] Since silica gels have been in use for direct encapsulation of living cells it holds a lot much of promise in terms of biocompatibility, in the form of silica-based bionanocomposites. The tremendous versatility of silica is responsible for its potential in a host of applications, on combining with different biocatalysts and biopolymers.

9.4   CARBON NANOMATERIALS

Carbon-based nanomaterials are vital constituents for designing a broad range of bionanocomposites. Since their advent, CNT, carbon nanofibers, graphene, graphene oxide (GO), and reduced graphene oxide (rGO) and their derivatives have been widely investigated for the configuration of bionanocomposites.[87–89] They possess excellent mechanical properties because of which they find applications as reinforcements in polymeric matrices.[90–92] Carbon-based materials impart electrical conductivity to composites.[93,94] Further, it minimizes the electrical percolation threshold in reference polymers and promotes hydrophilicity. An approach demanding the oxidation of the CNTs or graphene sheets to produce comprehensive oxygen-containing function at the particles' interface is most sought after for this purpose. The characteristic structure of the carbonaceous particles can facilitate mechanical dispersion in aqueous media and provide the materials with enough reactive sites to showcase the covalent structure of the carbonaceous materials. The ability to associate the CNTs with biomolecules has made it possible to develop CNT-based biosensors wherein the electric conductivity is used to advantage.[95,96] Similar approach was used in the later times to give an impetus to the colloidal stability of oxidized CNTs by grafting hydrosoluble polymers at the CNTs surface.[97,98] Further, successful use of graphene is attributed to the use of chemistry of CNTs for managing the exfoliation and stabilization of discrete graphene sheets. Moreover, the potential to oxidize graphene to GO and reduction to obtain rGO has led to development of highly conductive, possessing higher specific surface area materials as electrodes.[97,99–101] The applications of such family of materials are naturally related to bionanocomposites wherein electrical conductivity is instrumental for using it in microbial fuel cells (MFC)[102] or electrochemically active materials.[103,104]

9.5 KEY BIOCATALYSTS

Biocatalysts are used for their catalytic activity and found in living organisms initially. Biocatalysts can be whole cells and microorganisms other than biomolecules. The word biomolecule itself covers a wide spectrum of species. Different kinds of building blocks used in designing biocatalytic devices are biomolecules such as lipids or fatty acids, nucleic acids, polypeptides or proteins, and carbohydrates. It is like simplifying the big complexity of living organisms which can facilitate in investigation of the host of potential biocatalysts. Lipids and carbohydrates are not presently used straightway as biocatalyst but as structural components of biomaterials. Lipids are commonly used for the encapsulation of drugs in liposomes for delivering at localized sites. Polysaccharides are used most as the organic component in bionanocomposites and are called as structural and not functional building blocks. The line of demarcation between functional and structural building blocks may not be very clear. A system consisting of an immobilized biomolecule or whole cell embedded in an inorganic matrix is in itself a bionanocomposite or biohybrid material.[105] However, a bionanocomposite-based catalytic device is a matrix consisting of an inorganic and a bioorganic material, wherein a biomolecule or a cell is incorporated to impart the catalytic activity. Nucleic acids, like DNA or

RNA, and polypeptides, oligopeptides to larger proteins, have been fully articulated as biocatalytic entities in bionanocomposite materials. Few of the uses of biomolecules are biocatalysts, starting with the smallest and simplest molecules such as vitamins, hormones, and some kinds of neurotransmitters toward highly complex structures and the use of whole cells. Such small molecules although cannot be used as biocatalysts however can be very well used as cocatalysts, substrates, or products of the catalyzed process even.[106]

9.6   NUCLEOTIDES AND AMINO ACIDS

Nucleotides and nucleic acids lay the foundation for biomolecules hierarchy. Although the monomers (nucleotides) are not put to use in the conception of bioactive materials, the resulting polymers, such as synthetically derived nucleic acids, RNA or DNA, are extensively used in innumerable applications. In fact, DNA possesses quite exciting properties so as to use it as nanomaterials in different applications.[107] Many interesting applications of this wonderful structure, as building block for building nanoarchitectures and nanomachines, are due to its typical pairing properties. Its binding properties too can be used to advantage to construct biosensors and detection devices.[108] Further advantage of it is taken by entrapment of DNA strands in multiwalled carbon nanotubes (MWCNTs) to get bionanocomposite. The signal derived from this electrode indicated the DNA state, thus paving the way for an efficient detection technique for damages if any in DNA.[109] An iron oxide–chitosan electrode was designed with DNA to detect pyrethroid, which is used as an insecticide. These two examples exhibit the application of DNA due to its binding properties which can be studied using characterization techniques like UV–vis and FTIR spectroscopy. They won't find use as catalysts, as there is no change in a given chemical substrate. However, DNA strand in itself can be used as literal catalysts[107] as DNAzymes, or DNA sequence having catalytic activity, particularly in RNA cleavage, ligation,[110] amide and ester formation or Diels–Alder reaction. In fact, DNAzymes do not have structures which are occurring naturally but are produced by selection processes. The major application area of DNA-based materials is as biosensors. DNAzymes are employed as catalysts of the H_2O_2-mediated oxidation of 2,2-azinobis(3-ethylbenzothiazoline)-6-sulfonic acid because of the characteristic structure of DNA aptamers linked to hemin groups.

This reaction brings about a color change of the substrate and is therefore suitable for a sensor. By identifying the sequence of the DNA with care, targeting a compound of interest is possible. An adenosine triphosphate sensor was designed using such technique,[111] while the same structure with a different aptamer was used to sense thrombin.[112] DNAzymes are however not used as widely as protein-based catalysts. It is attributed to the technology which is quite new as against enzyme catalysis, which has been in use since long. Another limitation is DNA sequences contain handful functional groups, which does not allow efficient binding sometimes.[113] Amino acid-based biocatalysts may be a better substitute, given their widespread sequence of monomers. Indeed, for DNA and protein sequencing 4 nucleic bases and 21 amino acids are available, respectively. Moreover, these amino acids may be selective following translational modifications, producing infinite number of combinations as possibilities. Amino-acid-based molecules can vary from polypeptides to proteins possessing too short sequences and giant and complex configurations, respectively.[114] Simple amino acids such as L-alanine, L-glutamic acid, L-serine, and L-lysine in alginic acid/silica composites were used, with a platinum chelate for catalyzing the hydrogenation reaction of furfuryl alcohol. Percentages of 84.5 and 98.2 of product and optical yield were obtained on using glutamic acid-based catalyst. This limits one of the major merits of biocatalysts, which is to work as a substitute for traditional catalysts. There are many reports of using further complex amino acids, or proteins, to conceive bioactive materials. However, it should always be kept in mind all proteins cater to the needs of biocatalytic applications.

There are few proteins that contain a structural function, like collagen, keratin, or actin, which is instrumental in the conception of biocomposite material but as a building block structurally[115] and not functionally. However, some proteins are typically configured for cell indication, for example, antibodies which make them right candidates as biosensors. Antibodies have been extensively in use particularly in enzyme-linked immunosorbent assay (ELISA) test. But these analytical techniques take many steps, involving alternate addition of reagent and washing steps. In ELISA, an enzyme suspension is developed for catalyzing a colorimetric-based reaction. The antibody immobilization or encapsulation in biocomposites are used to advantage for providing an opportunity to incorporate all in a sensor, to derive a potentiometric signal, without reagent addition.[116] Antirabbit immunoglobulin G (IgG) was encapsulated in a polypyrrole propylic acid/TiO_2 nanowire film mounted on a gold microelectrode. Although this device does not contain a biopolymer, still it shows the potential of antibody encapsulation, to sense rabbit IgG with a linearity varying between 11.3 and 113 µg/ml. Antirabbit IgG was immobilized to sense secondary antibody, that is, the rabbit IgG. There are many examples of antibody immobilization to detect various substrates.[117] Rabbit IgG is immobilized in an iron oxide–chitosan bionanocomposite film coated on indium tin oxide electrode to detect ochratoxin-A, a mycotoxin in food-stuff at a concentration of 0.6 ng/dl. Few proteins offer binding sites and act as carriers of small molecules, for example, hemoglobin, which carries oxygen all through the body of an organism.[118] Successful encapsulation of hemoglobin in a chitosan/$CaCo_3$ matrix was done and thereafter deposited on a glassy carbon electrode, to bring about catalytic reduction of H_2O_2. The microelectrode therefore was used as an H_2O_2 biosensor.[119]

Therefore, it is a fact that the high-binding tendency and specificity of small biomolecules are of great importance to design the bioactive materials which make them right candidates for specific applications, particularly as biosensors. However, biomolecules and proteins find industrial applications too typically as catalysts.

9.7   ENZYMES

Widespread works have been undertaken on application of enzymes as biocatalysts.[120] Enzymes are proteins which are specific in action in the organism as catalysts. The efficiency of them is attributed to the presence of one or more binding sites which can fix the given substrate(s) and releasing the target product(s). These biomolecules are part of almost every metabolic step and are therefore many thousands in number. Enzymes have been categorized into various types based on the kind of reaction they catalyze such as oxidoreductases, transferases, hydrolases, lyases, isomerases, and ligases. The enzymes are almost the same in number as substrates. The use of enzymes as catalysts on industrial-scale demand confirming the requirements of ecology and sustainability. Enzymes were considered as ecobenign substitutes to toxic reagents and solvents.[120] Since then, the application of enzyme on industrial scale gained momentum and is the most sought after in diverse areas such as food and brewing industry, fuel reformulation, biodetergents, biodegradation, paper, textile, and so on. Innumerable enzymes are used for various applications but some are more preferred over others.[121] Different uses of lipases (triglycerides hydrolyzing enzymes) have been reviewed by some researchers in several industrial domains. The enzyme finds use in bioreactor but is used as bioactive material part, either by entrapment or immobilization, and focus on limited applications is of interest. Biomolecules in general have the ability to bind specifically to their targeted substrate, to be used as efficient catalysts for biosensing purposes. The organic and inorganic species are very diverse, bionanocomposites, chitosan, and alginate-based materials are of great interest. The inorganic substances used are varying between graphene and CNTs to moieties obtained by sol–gel method, using clays and LDH.[122] There were works on a phenolic compound sensor with the help of tyrosinase,[123] a cholesterol sensor using enzyme, cholesterol oxidase,[124] or entrapment of enzyme glucose oxidase for the detection of glucose. Such examples prove the versatility for the use of enzymes and their functionality worth tuning. In a nutshell, the potential of the enzyme-based approach lies in its versatility and specificity, leading to the potential in finding a right enzyme target pair for varied applications. However, the approach has its own limitations too when, for example, proteins and biomolecules, in general, have stability issues. Proteins have 3D structures but fragile. A marginal modification in their structure can substantially suppress the catalytic activity. Various parameters are responsible for such kinds of structural modifications, typically environmental factors like temperature or pH. Enzymes are vulnerable to inhibition mechanisms too, leading to deactivation of the biocatalysts. There are some approaches to get over these limitations, such as immobilization of the enzyme so as to ensure more stability. Immobilization can be accomplished by means of various techniques like physical binding to a material with the help of hydrophobic, van der Waals, or ionic or covalent interactions and by entrapping in a target matrix.[125] Approaches need to be developed amenable to entrap in inorganic matrices,[79] but emphasis is laid upon immobilization within composite materials and particularly biocomposites so as to obtain optimum biocompatibility.[126] Thermal and chemical stability of β-glucosidase has been reported on immobilization within the beads of silica–alginate. Biomolecules have been in use as quite versatile and efficient biocatalysts, catering to the needs of host of applications.

Another strategy to bioinspired catalysis is use of a few selected molecules derived from biological organisms or whole cells to take advantage by mimicking the mechanics of nature.

9.8   WHOLE CELLS

Natural resources can be used as catalysts, directly as whole cells, in place of specific enzymes or other biomolecules. Although a cell is a quite complex, it still is an efficient architecture, and therefore using only a part of it without compromising performances is a huge challenge. There are indeed limitations in using living cells, but the benefits are too many. In regard to whole cells as biocatalysts, a host of cell types is worth considering. For example, mammalian cells are only part of a larger living organism. They are not much used as biocatalysts as it is but instead as carriers in medical applications. Plant cells too have been in use as functional building blocks. Similar to mammalian cells, plant cells too are a part of larger complex organisms, but carefully selecting cell type decides performance efficiency in typical applications. There have been efforts of encapsulation or immobilization of simple organisms like algae that can be either unicellular or multicellular. Moreover, fungi varying from simple unicellular such as yeast and mold to larger multicellular microorganisms have been entrapped for numerous applications. Majority cells are eukaryotes that contain nuclei and different organelles. On the other hand, prokaryotic cells like bacteria that are bereft of both nuclei and organelles and are normally unicellular species found across the board right from soil to human body. Mammalian cells are preferentially used for encapsulation[29] and cell delivery[127] or as part of synthetic scaffolds.[115] However, they find restricted use in medical applications but not as biocatalysts. Plant cells are commonly used for biocatalytic applications.

The first and foremost use of plant cell culture is as "factories"[128,129] for metabolites generation. Of course, several plants generate chemical species characterized by pharmaceutical or medicinal properties, and employing the plant cells directly to produce such metabolites in an economical and eco-benign manner. Some plants have the ability of photosynthesis that metabolize CO_2 into carbohydrates, which is of great interest for energy-based applications. Hence, identifying and encapsulating plant cells are the right approach to configure leaf-like materials amenable to produce carbohydrates.[130] Plant cells find use as part of biosensors.[131] Tobacco cells have been used for the detection for flagellin, to detect pathogenic bacteria. However, most whole-cell-related sensors are based on other cellular types. The use of isolated plant cell for bioremediation is comparatively rare. Of course, bioremediation via plants takes place by virtue of phytoremediation[132] which allows for the use of whole plants and not isolated cells for degradation of contaminants.[133] Tobacco cells have been used to degrade 2,4,6-trinitrotoluene. In this regard, the cells do not belong to bionanocomposite device but present as cellular suspension. Algae too find use because of their photosynthetic ability. Of course, algae find extensive use in the generation of many chemicals using CO_2, such as biodiesel and biohydrogen.[134] Here, too, the cells will not find much use as catalysts but instead as complete factories. Uses of algae as a catalyst in various chemical reactions namely reduction of acetophenone derivatives have been reported by earlier researchers.[135] The biocatalytic activity of algae is not restricted to the synthesis of typical chemicals. Of course, the biocatalytic process is of great interest for designing biosensors.[136] Microalgae *Chlorella vulgaris* entrapped in a sodium alginate/CNT matrix were used and thereafter coated on a carbon electrode to detect numerous pollutants, like atrazine, wherein the oxygen produced by the microalgae is instrumental in detection. Algae have proven potential in the treatment of wastewater contaminated with nitrate and phosphates salts[137] as well as degradation of bisphenol-A and benzophenone.[138] Fungi too can be used for bioremediation purposes. For example, white rot fungi, which generates host of enzymes bringing about degradation of numerous chemicals efficiently, from lignocellulosic residues[139] to polychlorinated biphenyls (PCB), polycyclic aromatic hydrocarbons (PAH), and so on.[140] Smaller fungal species can be used as functional species in biosensors,[141] for example, yeast and molds have been in use too for such applications. Fungal species too can be used in the synthesis of chemicals similar to plant cells and algae. For example, *S. cerevisiae* is a very commonly and extensively used yeast (called as baker's yeast) found to reducing agent for different chemicals.[135,142] Fungi and yeast exhibit biocatalytic activity for the cause of biodiesel production.

All such eukaryotic microorganisms have proven potential as effective biocatalysts, although it is not much popular option. Since bacterial species are of course very diverse such that they can perform variety of catalytic activities, it is not possible to quote an exhaustive list of the various bacteria that are used for each kind of application. Enzymes are very common in use for industrial-scale biocatalytic applications in the production of chemicals. A broad range of enzymes is used in these processes right from microbial origin, in the form of whole microbial cells as catalysts, for example, ketone reduction, C–H hydroxylation, nitrile hydrolysis,[135] and so on.[143] However, relatively very few selective species of bacteria like cyanobacteria are photosynthetic as against plant cells or algae, although few applications on energy production front rely upon bacteria.[144] *Pseudomonas* and *Shewanella* species have been reported to be used as functional component of MFC for hydrogen and electricity production.[145] The versatility of bacteria can be used to advantage in biosensing, transducers in the configuration of electrochemical, or optical sensing devices.[146] Such biosensors are used preferentially for the detection of environmental contaminants,[147] as against biomolecule-based ones which are frequently used as biomedical sensors. The application of microbial biomaterials in the environmental sector is not only restricted to detection devices but as functional component of materials used for bioremediation of nitroaromatic compounds,[148] heavy metals,[149] dyes,[150] PAH,[151] or PCB apart from other targeted compounds.[152]

Whole-cell catalytic systems are varied a lot based on cell types and are favored for environmental applications such as energy production or contaminant detection and removal. The use of whole cells has comparative advantages over simpler biomolecules are innumerable, cost-effectiveness. Biomolecules, namely, enzymes and antibodies are often too expensive to generate and hence are not often an economically viable option for industrial-scale applications. Whole cells in this regard are much easy to handle and grow, subject to monitoring a few factors like temperature or pH. Such cells find place biocatalytic reactors, naturally designed and characterized by complexity and efficiency which are not very easy to obtain using artificial devices. However, there are some limitations too while using whole cells that are living organisms and are in particular fragile and sensitive to the environment to which they are exposed. It can provide valuable toxicity-related information of their environment using a biosensing device in terms of stability and durability. Although these living organisms grow to an advantage, they have a limitation in the form of difficulty in controlling the cell density precisely over time. These limitations however can be reduced by either encapsulating or immobilizing the cells in suitable matrices,[153] offering a physical barrier to growth, as well as against potential hazardous environment. All such examples exhibit the diversity of potential biocatalysts, as biomolecules (DNA, proteins, enzymes) or whole cells (plant cells, algae, fungi, or bacteria), paving the way for the design of largely specific and efficient biocatalytic devices. The application frontiers vary from biosensing, bioremediation to catalysis of industrial-scale chemical processes or energy generation. Biocatalysts are selective inaction for specific kind of applications. For example, bacteria are extensively used for remediation of environmental pollution but not well suited for biomedical devices. Similarly, certain proteins are efficient in catalyzing selective reaction for biosensing application although not useful in catalysis for industrial-scale production. The disadvantage of all such biocatalysts is their stability quite often, but this problem can be overcome by either immobilization or encapsulation of the given biocatalyst in a target matrix. The said matrix needs to be designed well for optimizing the stability and efficiency of biocatalyst. The application of a bionanocomposite is a highly strategic option which couples biocompatibility and structural properties. The key to design highly efficient biocatalytic devices lies in the choice of a suitable biocatalyst as functional building blocks, in accordance with the expected application, and structural building blocks to optimize the efficiency of the species on encapsulation.

9.9   APPLICATIONS OF BIONANOCOMPOSITE MATERIALS

The present state of the art is in assembling both functional and structural units to construct bionanocomposites with capability to effectively catalyze various biochemical reactions since the domain is vast and complex. The most marked examples that transform host of approaches, and diversity of application domains wherein bionanocomposite materials integrated with biocatalysts, are considered. Major applications of bionanocomposites are in biosynthesis, sensing devices, energy generation, and bioremediation. Following is an overview of present state of these referred applications, and the major approaches and materials employed for each of them are discussed.

9.10   BIOSYNTHESIS

The ability to hasten biochemical reactions under comparatively milder conditions is an important aspect in biocatalysis as it can produce desired compounds in a sustainable, economical, viable, and also expeditiously. However, many biocatalysts tend to denaturate, and thereby inactivate. For example, enzymes are typically sensitive to the environmental factors such as pH, temperature, and so on. To get over these limitations of enzymes as biocatalysts depends upon how they are encapsulated in bionanocomposite materials. LDH are found to be better enzyme hosts, for example, α-amylase entrapment by Mg–Al LDH is used as a starch biocatalyst,[154] while transketolase in Zn–Al and Mg–Al LDH[155] produces better yield in ketose synthesis. Clay minerals too are used for enzymes immobilization, namely, invertase, alkaline phosphatase, β-glucosidase, or lipase with quite encouraging findings as regards their activity and stability.[156–160] When enzymes were encapsulated by clay minerals and LDH, they were immobilized in the inorganic material, without any support of a biopolymer species.[161,162] Grafting with covalent interactions of the enzymes to alginate–montmorillonite beads through glutaraldehyde cross-linking was also proposed and found that the highest rate of reaction with immobilized enzymes was far more than with free enzymes.[161]

One more very common strategy to improve the activity and stability of enzymatic moieties is to entrap it in alginate–silica beads.[163] Industrially important reaction of starch hydrolysis was focused upon and elaborated these materials with the help of simple mixing of silica and alginate before the formation of beads.[85] Nanocomposite architecture has a substantial effect on the enzyme stability as it has caused the enzyme use in 20 starch hydrolysis operations in succession, to reduce the starch hydrolysis reaction cost.[85] One more architecture using alginate beads has also been proposed[164] and alginate core–silica shell structure was derived by preparing alginate beads cross-linked with calcium, followed by successive coating of poly-L-lysine and silica. The alginate–silica beads were then treated with citrate ions which acted as a calcium chelating agent, thereby the inner alginate core was left liquefied. The catalytic activity of β-galactosidase for the degradation of 4-nitrophenyl-β-D-galactopyranoside was marginally lower than that of the free enzyme, the beads could be washed and reused exhibiting catalytic activity in semblance.[165] An alternative morphology with entrapment of *Chromobacterium viscosum* lipases in gelatin–silica hardened gels was obtained. The device so resulted catalyzed trans-esterification of fatty acids into cyclohexane quite efficiently, with 85% product recovery.[166] Similarly, *Candida rugosa* lipases were encapsulated in a biopolymer/silica gel for applications of hydrolysis. The biopolymer chosen was sporopollenin, found in spore membranes, a very potential candidate for encapsulation as it tends to form small cavities. The device exhibited encouraging results, typically in the form of enantioselectivity. The typical use of silica provided for sound mechanical properties as against biopolymer in isolation, prevented leaching of the encapsulate, and presenting a better storage and thermal stability in these examples.

Few of the latest examples on the bionanocomposites frontier show biocatalytic activity with emphasis on the design of materials that can handle key problems in enzymatic catalysis over and above biocatalytic activity and stability. Of course, recovery and reusability of the catalyst are key issues in designing cost-effective and efficient processes. To address these problems bionanocomposite materials integrate magnetic particles with the help of an external magnetic field which will provide for the material recovery.[167] Immobilization of bovine α-chymotrypsin (αCT) on silica-coated super-paramagnetic nanoparticles, Fe_3O_4@silica was proposed. The enzyme is covalently bonded to the silica nanoparticles surface by using glutaraldehyde crosslinks between the enzyme and the amino groups created earlier by grafting 3-aminopropyltriethoxysilane on the silica surface. The hybrid (Fe_3O_4@silica-αCT) so resulted showed stability in its activity toward Z-Ala-Phe-OMe peptide by ester hydrolysis. Moreover, the nanoparticles were recovered in full by applying magnetic field externally.[168] Improved catalytic activity and stability for more complex architectures were reported. For example, nanosilica coated with nanocobalt ferrite were mixed with oxidized CNTs. Immobilization of laccase using covalent 1-ethyl-3-(3-dimethylaminopropyl)-carbodiimide (EDC)-bonded cross-links on the oxidized CNT walls was done. On comparing with the free enzyme, the catalytic activity of immobilized laccase on the bionanocomposite was kept intact to a large degree at temperature varying from 50 to 70°C and pH ranging between 5 and 8. The stability observed is relevant keeping in view the parameters governing process under test are in tandem with the operational limits of the enzyme in question.[168]

One more example related to the catalytic activity of encapsulated enzymes in bionanocomposite materials regarding the use of glucose oxidase and horseradish peroxidase together to stimulate the enzyme-catalyzed polymerization of poly(ethylene glycol)methacrylate (PEGMA) in a composite material with layers.[169] It relies upon the dispersion and piling up nanosheets of calcium niobate along with two enzymes to produce a hybrid material containing layers. The ultimate material was then dispersed in water and PEGMA where the two enzymes based underlying cascade mechanism catalyzed the PEGMA polymerization followed by exfoliation of the layers of calcium niobate.

9.11 BIOSENSING

The most vital application of the biocatalytic systems is designing typical sensing devices. Most importantly enzymes have been greatly used to advantage for the development of amperometric biosensors, for example, use of enzyme, horseradish peroxidase for biosensing of H_2O_2. There are many experimentations pertaining to this enzyme in particular, with changes in the biopolymer used or the configuration of the device. A widely used biopolymer for such applications is chitosan, where a hybrid chitosan/silica film is mounted on a given electrode like carbon paste electrode[170] or platinum.[171] The efficiency depends upon the device, in the form of detection limits ranging from 8.2×10^{-8} to 3.1×10^{-6} M and linear response range varying between 2.1×10^{-7} and 3.5×10^{-3} M. In this case, inorganic materials that do not possess electronic conductivity namely clay minerals, silica, or LDH are normally mixed with conductors like Au nanoparticles,[171] metal oxides,[172] or MWCNTs.[173] Though horseradish peroxidase is well known for sensors,[171,172] other enzymes namely cholesterol oxidase[123] and cholesterol esterase[174] have been in use by immobilization on chitosan/silica films to detect cholesterol for medical applications. The drawback of such applications is the electrical insulating property of silica, which is made up by the addition of MWCNT. Similar to silica, LDHs have also been largely exploited as the inorganic species to design enzyme host for biosensors.

The direct immobilization of enzymes in LDH for electrochemical sensing has been widely explored.[175,176] More complex materials, obtained by combination of knowledge and the advantages derived by the mixing of a biopolymer species to the inorganic-based LDH matrices, molecular nonenzymatic catalysts are rarely employed to design biocatalytic devices.[177] The intercalation of hemoglobin in a Ni–Al LDH, along with DNA strands take place. The LDH offered protection from protein denaturation, while DNA showed a synergetic effect with hemoglobin, producing a device with excellent electrocatalytic characteristics.

Another very commonly used model enzyme in catalytic sensing is glucose oxidase. The immobilization of this enzyme in alginate/Zn–Al LDH[178] or in chitosan/Zn–Al LDH[124] is done. However, entrapping the enzymes offered better pH and thermal stability to the devices while generating excellent sensing efficiencies in the form of activity range and sensing limits. The biopolymer added proved invaluable as it bettered the mechanical properties of the films, particularly swelling and cracking when stored. Similarly, cholesterol oxidase entrapped in Zn–Al LDH[179] has been in use as a cholesterol biosensing device with high degree of efficiency. The performances of such device were improved by coencapsulating cholesterol oxidase and horseradish peroxidase, generating a good sensing device with better detection efficiency, long-term operational stability, and greater specificity toward its target. Polyphenol oxidase has been entrapped too in chitosan/Zn–Al LDH[180] and alginate/Zn–Al LDH[181] composite materials matrices. The resultant sensor showed sound detection efficiencies toward its target (catechol), and thermal, operational, and storage stability.[118] Entrapment of hemoglobin in chitosan films having calcium carbonate nanoparticles was explored too. Such films were used to alter carbon glassy electrodes so as to design an H_2O_2 sensor free from redox mediator. This is attributed to the good electron transport characteristics of the hemoglobin-based system and $CaCO_3$ nanoparticles, over and above their structural characteristics, substantially improved the sensing device's performances. The reason being the nanoparticles may bring about an orientation precisely of the protein, producing an optimization of the electron transport between the protein and the carbon glass electrode. Therefore, structural and functional building blocks intimately interact and the choice of the structural biopolymer and inorganic species matter a lot. Detection of target analytes by bionanocomposites depends upon biocatalysts, and the use of living bacterial cells supports the detection of analytes of interest.[182] For example, alginate was coated on a carbon led to highest hydrophilicity. The newly developed hydrophilic carbon was incubated with *Escherichia coli* and subsequently by coating of silica to create encapsulation of bacterium in a 3D mesoporous carbon silica composite. In presence of glucose in solution, the metabolic activity of the bacterium resulted into a detectable electrochemical signal indicates the presence of nutrients.

9.12   ENVIRONMENTAL REMEDIATION

Applications of bionanocomposites for bioremediation demand the use of whole cells as functional biocatalysts. The term "whole cell" involves a host of species ranging from bacteria to algae, fungi, as well as plant cells.[183] Even mammalian cells can be encapsulated limited however to medical applications. Such a kind of encapsulation asks for a biocompatible matrix, which can act as a barrier to potentially hazardous environment a material possessing requisite mechanical properties. Silica is considered as a promising candidate for such kind of application.

For enzymes, a classical host configuration is the alginate/silica core–shell bead.[184] *Nostoc calcicola*, a bacterium, was entrapped successfully in such a bead structure and checked the activity of the device as a biosorbent toward heavy metal ions such as Cd^{2+}, Cu^{2+}, Cr^{3+}, and Ni^{2+}. The alginate/silica composites too have been used as hosts for fungi[185] and for the remediation of malachite green. The bead structure is however not enough for scaled-up applications.[186] Entrapment of *E. coli* in electrospun PVA–silica fibers was also proposed which can degrade atrazine. PVA however is not a biopolymer but a water-soluble synthetic polymer. We use biopolymer to produce a biocomposite and can enhance the biocompatibility of the matrix to entrap a host of metabolically active bacterial species. The use of whole cells is not confined only to bioremediation. Although conventionally enzymes are used in preference for biosensing,[187] immobilizing *C. vulgaris*, a microalga, within alginate beads, and subsequently entrapping it in a silica matrix can open up new application avenues. Transformation of fluorescein diacetate into fluorescein is catalyzed by esterase obtained from these algae. The enzyme generation has been inhibited by pesticides, hence such device can straightway be used for biosensing purpose. For such typical application, silica is used as the inorganic species. It is possible to obtain a transparent silica coating under certain conditions, and thereby providing a protective barrier while detecting fluorescence. LDH-based bionanocomposite materials can be of vital importance in the area of bioremediation. LDHs are already in use, because of their characteristic structure that produces quite interesting adsorption properties toward numerous pollutants.[188] An alginate–LDH composite adsorb Orange II dye efficiently than its separate components. This is however not called bioremediation, but simple removal of a pollutant from a localized site. It is therefore of great interest to directly bring about the degradation of the pollutant, because of the activity of biological entities. To accomplish this objective, it is necessary that encapsulation of bioactive species in such bionanocomposite materials is a must. As discussed earlier, in bioremediation the well-known and highly efficient biocatalysts are whole cells, specifically bacteria. *Pseudomonas* sp. strain adenosine di-phosphate (ADP) was immobilized in Mg–Al LDH using intercalation of the cell in between LDH layers[189] or direct coprecipitation bacteria are present.[105] Both strategies yield a bionanocomposite material which degrades atrazine efficiently, with improved activity and reusability as against free cells.[105] Although simultaneous use of both strategies such as use of LDH-based bionanocomposites and encapsulation of species degrading pollutants is not reported so far, but it is beyond doubt that it would be of great interest to use to advantage the improved adsorption capabilities of LDH-based hybrids coupled with the degradation capabilities of encapsulated species to develop highly efficient in-situ devices for bioremediation. Bioremediation processes normally favor the use of bacteria instead of mere enzymes, as the latter ones are costly, sensitive, and difficult to recover for reuse. However, the remedy is encapsulating it within a bionanocomposite which allows for exploiting the wide versatility of enzymes to combat pollution problems. For example, lipases can be used to degrade fatty residues in wastewater or convert grease waste into biodiesel.[190] Laccases are oxidases containing copper centers and have been used in many bioremediation of many organic pollutants, for example, synthetic dyes, trichlorophenols, bisphenol A, alkenes, or herbicides.[191] One more example of enzyme employed for remediation is the peroxidase[192] to degrade PAH, organophosphorus pesticides, and phenols.[193] Peroxidases encapsulated in alginate/silica particles were used for remediation of phenol. The underlying principle used is that peroxidases have ability of catalyzing phenol polymerization. About 85% conversion was achieved compared to 70% for free enzymes. A regeneration can substantially enhance the reusability of the bionanocomposite.[193]

9.13 ENERGY GENERATION

One of the key application areas for biocatalytic materials is fuel cells. In preference, MFCs are used, wherein bacteria act as catalysts for oxidizing organic and inorganic materials and produce current.[194] MFCs are examples showing synergy between the bionanocomposites and biocatalytic species. A characteristic feature such materials possesses is the formation of biofilms at the surface or within the pores of electrodes. Biofilms are bacterial colonies, self-generated, polysaccharide ECM. The coupling of the polysaccharides is generated by bacteria when biofilms form in presence of conductive fillers as particulates that account for the development of bionanocomposite material devices having biocatalytic activity with potential applications in bioremediation of organic compounds and energy generation. Other biocatalytic devices instrumental in energy generation are entrapped algae-based ones.[195–197] Similar to cell-based fuel cells, other biocatalytic devices amenable to the generation of energy namely fuel cell-based enzyme has been suggested.[198] Various types of strategies employed, the hydrophilicity of the materials that are conductive is often facilitated by the coating of biopolymers like alginate or chitosan to improve upon the intimacy between the biocatalytic and the conductive moieties.[86,199] These devices can be used in different areas particularly in environmental remediation, energy generation, and fuel production. Conductive carbon nanomaterials notably graphene and CNT are most sought after materials in the assembly and efficiency of MFC.[200] The better electronic conductivity and stability of colloidal solution of graphene and GO, respectively, have been reported to be beneficial for designing MFC. An emphasis has been laid upon the microbial respiration in anaerobic activated sludge biofilm to reduce GO to graphene using glucose following which the anode chamber electrode was incorporated into the cathode chamber and made use of as the graphene biocathode in a MFC involving reduction of Cr(VI). As against a graphite biocathode, the graphene-embedded electrode enhanced reduction rate of Cr(VI) accompanied by five times rise in rate constant to 0.169 h^{-1}. Simultaneously, the MFC electricity generation raised from 28.7 to 163.9 mW m^{-2}.[200] The marked rise in efficiency of the graphene-reinforced electrode was attributed to the sudden rise in specific surface area of graphene as compared with carbon felt, along with a reasonable rise in the electrical conductivity. However, the morphology of graphene influences the production of a conducive 3D environment to entrap cells. To get over the morphological and electron transport-related drawbacks,[201] a bionanocomposite wherein the connection of rGO to MWCNTs helps to circumvent such limitations. A *Shewanella putrefaciens* biofilm was used as the anode in an H type double chamber MFC and the MWCNT@rGO1:2/CC anode produces a highest power density of 790 mW m^{-2}, far more than MWCNT/CC (424 mW m^{-2}) and rGO/CC (501 mW m^{-2}).[201] There has been an important role of the composite structure and its effect on the efficiency of biocatalytic systems for energy generation. While the biocatalytic systems aimed at producing electricity are largely based upon the incorporation of conductive materials like carbon felt electrodes, metal nanoparticles, conductive carbon-related nanomaterials, or fuel cells derived based upon photosynthesis process have far different demands. In these cases, the transparency of the parts stands is the key property. Silica has been preferably used to advantage[202] to design photosynthetic solar cells. For example, encapsulation of *Dunaliella tertiolecta*, a microalga with high photosynthetic activity in alginate/silica beads, was achieved, so as to generate oxygen for 25 days.

The use of microalgae in place of bacterial biofilms for the cause of energy generation is a general trend that is observed from the earlier research.[145,195–197,203] Apart from the use of whole cell-based systems, enzyme-based fuel cells have been used for energy generation. Indeed, enzymes too can be instrumental in the configuration of eco-benign energy systems and in particular biofuel cells.[204] Various types of enzyme-based biofuel cells and some relevant characterization techniques have been reviewed. Enzymes from the bioreductase class used in biofuel cell are preferred, and the most common example is glucose oxidase.[86] Enzyme glucose oxidase encapsulated in alginate/carbon beads was used as a promising candidate for biofuel cells. Other researchers have employed various enzymes for the configuration of these kinds of cells.[205] A laccase based biocathode for the reduction of di-oxygen as well as an alcohol dehydrogenase-based bioanode for the oxidation of ethanol was used too.[205] A biofuel cell was designed containing alcohol dehydrogenase as the anode and laccase as the cathode and in presence of ethanol (e.g., alcoholic beverages) for the generation of power density and current density of the order of 1.56 mW cm^{-2} and 2.08 mA cm^{-2}, respectively. Further, the device setup demands the conductive species for the conduction of electric current. It was proposed by the earlier researchers that carbon felt was a basis material in the making of the electrodes followed by doping with gold nanoparticles was essential for achieving better performance.[206,207]

9.14 CONCLUSION

One of the very vital issues on the bionanocomposite materials frontier for biocatalytic applications is closely related to structural diversity. This family of materials comprises of wide spectrum of catalysts varying between biomolecular catalytic species like hemoglobin and living cells. Simultaneously, the structural species are also quite diverse in nature, ranging between CNTs and collagen. Such a kind of diversity gives rise to broad range of biocatalysts related to bionanocomposite systems. Indeed, this compositional diversity is an asset for designing materials on demand which can be maneuvered into complex structures to deal with typical biocatalytic challenges. However, the mentioned structural diversity means a broad and heterogeneous research fraternity is involved in addressing the same issue having different scientific backgrounds, leading to numerous expectations and even a different lexicon, providing for generation of a varied and difficulty in comparing the literature in the domain. However, the literature in this area is very vast. Although not exhaustive, the selected works in this chapter will provide the reader with a broader perspective of kind of trends that have come up over the time in this field. Irrespective of the domain from sensors to energy generation, the scientific works are aimed at proof of concepts which are still yet to be implementation on a greater scale. One of the reasons for nonimplementation on a larger scale may be the regulatory norms regarding nanomaterials on the industrial front, largely because of nanotoxicity issues.

The bionanocomposites are used for improvement of biocatalytic activity reflected upon efficiency. Bionanocomposite materials find use as biocatalysts that are known for specificity although not much sensible variation takes place in it. However, biocatalysts in finely tuned architecture forms with chemical environments in congruence with their biocatalytic activity can largely boost the diffusibility of the substrates or mediators, which will finally result into rise in productivity. Similarly, the well-designed bionanocomposite architectures as the biocatalyst can promote increase in stability towards drastic physicochemical conditions that may affect the catalysts' efficiency. Most of the works described in this chapter depend upon conventional processing techniques to design the bionanocomposite and thus the biocatalyst. Some of the works however throw light upon innovative processing techniques such as electrospinning or bioprinting which can fine-tune living matter.[206,207] These innovative processing technologies could broaden the spectrum of applications of bionanocomposite materials on biocatalysis front.

KEYWORDS

- **bionanocomposites**
- **nanotoxicity**
- **biocatalytic activity**
- **electrospinning**
- **bioprinting**

REFERENCES

1. Sullivan, L. *Lippincott's Mag.* **1896**, *1*, 403–409.
2. Singh, B. K. *Nat. Rev. Microbiol.* **2009**, *7*, 156–164.
3. Puri, M.; Sharma, D.; Barrow, C. J. *Trends Biotechnol.* **2012**, *30*, 37–44.
4. Esser, K.; Lemke, P. A.; Bennett, J. W.; Osiewacz, H. D. *Ind. Eng. Chem.* **1925**, *17*, 445–450.
5. Bruggink, A.; Roos, E. C.; de Vroom, E. *Org. Process Res. Dev.* **1998**, *2*, 128–133.
6. Johnson, E. T.; Schmidt Dannert, C. *Trends Biotechnol.* **2008**, *26*, 682–689.
7. Ranganathan, S. V.; Narasimhan, S. L.; Muthukumar, K. *Bioresour. Technol.* **2008**, *99*, 3975–3981.
8. Casero, E.; Darder, M.; Pariente, F.; Lorenzo, E. *Anal. Chim. Acta* **2000**, *403*, 1–9.
9. Jeykumari, D. R. S.; Narayanan, S. S. *Carbon N. Y.* **2009**, *47*, 957–966.
10. Ayyub, O. B.; Kofinas, P. *ACS Nano.* **2015**, *9*, 8004–8011.
11. Ikeda, M.; Tanida, T.; Yoshii, T.; Kurotani, K.; Onogi, S.; Urayama, K.; Hamachi, I. *Nat. Chem.* **2014**, *6*, 511–518.
12. Mailloux, S.; Katz, E. *Biocatalysis* **2014**, *1*, 13–32.
13. Gerlt, J. A.; Babbitt, P. C. *Curr. Opin. Chem. Biol.* **2009**, *13*, 10–18.
14. Veitch, N. C. *Phytochemistry* **2004**, *65*, 249–259.
15. Rayu, S.; Karpouzas, D. G.; Singh, B. K. *Biodegradation* **2012**, *23*, 917–926.
16. Bornscheuer, U. T.; Huisman, G. W.; Kazlauskas, R. J.; Lutz, S.; Moore, J. C.; Robins, K. *Nature* **2012**, *485*, 185–194.
17. Darder, M.; Aranda, P.; Ruiz Hitzky, E. *Adv. Mater.* **2007**, *19*, 1309–1319.
18. Ruiz-Hitzky, E.; Aranda, P.; Darder, M. *Kirk Othmer Encyclopedia of Chemical Technology*; John Wiley & Sons, Inc.: Hoboken, NJ, 2008; pp 1–28.
19. Cooper, G. M. *Cell: A Molecular Approach*, 3rd ed.; Sinauer Associates: Boston, MA, 2000.
20. Caliari, S. R.; Burdick, J. A. *Nat. Methods* **2016**, *13*, 405–414.

21. Lim, F.; Sun, A. M. *Science* **1980,** *210,* 908–910.
22. Langlois, G.; Dusseault, J.; Bilodeau, S.; Tam, S. K.; Magassouba, D.; Hallé, J.-P. *Acta Biomater*. **2009,** *5,* 3433–3440.
23. Wang, N.; Adams, G.; Buttery, L.; Falcone, F. H.; Stolnik, S. J. *Biotechnology* **2009,** *144,* 304–312.
24. Christoph, S.; Kwiatoszynski, J.; Coradin, T.; Fernandes, F. M. *Macromol. Biosci.* **2016,** *16,* 182–187.
25. Zhong, M.; Sun, J.; Wei, D.; Zhu, Y.; Guo, L.; Wei, Q.; Fan, H.; Zhang, X. J. *Mater. Chem. B* **2014,** *2,* 6601–6610.
26. Hou, C.; Qi, Z.; Zhu, H. *Colloids Surf. B: Biointerfaces*. **2015,** *128,* 544–551.
27. Das, B.; Roy, A. P.; Bhattacharjee, S.; Chakraborty, S.; Bhattacharjee, C. *Ecotoxicol. Environ. Saf*. **2015,** *121,* 244–252.
28. Zhang, W.; Xu, F. *ACS Sustain. Chem. Eng*. **2015,** *3,* 2694–2703.
29. Uludag, H.; De Vos, P.; Tresco, P. A. *Adv. Drug Deliv. Rev*. **2000,** *42,* 29–64.
30. Champagne, C. P. *Immobilized Cell Technology in Food Processing*; Elsevier Masson SAS: Amsterdam, 1996; p 2.
31. Gasperini, L.; Mano, J. F.; Reis, R. L. *J. R. Soc. Interface* **2014,** *11,* 20140817.
32. Rocha, P. M.; Santo, V. E.; Gomes, M. E.; Reis, R. L.; Mano, J. F. J. *Bioact. Compat. Polym*. **2011,** *26,* 493–507.
33. Moslemy, P.; Guiot, S. R.; Neufeld, R. J. *Enzyme Microb. Technol*. **2002,** *30,* 10–18.
34. Moslemy, P.; Neufeld, R. J.; Guiot, S. R. *Biotechnol. Bioeng*. **2002,** *80,* 175–184.
35. Pereira, D. R.; Silva Correia, J.; Caridade, S. G.; Oliveira, J. T.; Sousa, R. A.; Salgado, A. J.; Oliveira, J. M.; Mano, J. F.; Sousa, N.; Reis, R. L. *Tissue Eng. Part C Methods* **2011***, 17,* 961–972.
36. Luan, N. M.; Iwata, H. *Biomaterials* **2012,** *33,* 8075–8081.
37. Suwandi, J. S.; Toes, R. E. M.; Nikolic, T.; Roep, B. O. *Clin. Exp. Rheumatol*. **2015,** *33,* 97–103.
38. Rinaudo, M. *Prog. Polym. Sci*. **2006,** *31,* 603–632.
39. Hunt, N. C.; Grover, L. M. *Biotechnol. Lett*. **2010,** *32,* 733–742.
40. Wong, C. T. S.; Foo, P.; Seok, J.; Mulyasasmita, W.; Parisi Amon, A.; Heilshorn, S. C. *PNAS* **2009,** *106,* 22067–22072.
41. Estrada, L. H.; Chu, S.; Champion, J. A. *J. Pharm. Sci*. **2014,** *103,* 1863–1871.
42. Lee, H. J.; Park, H. H.; Kim, J. A.; Park, J. H.; Ryu, J.; Choi, J.; Lee, J.; Rhee, W. J.; Park, T. H. *Biomaterials* **2014,** *35,* 1696–1704.
43. Picot, A.; Lacroix, C. *Int. Dairy J*. **2004,** *14,* 505–515.
44. Gunasekaran, S.; Ko, S.; Xiao, L. J. *Food Eng*. **2007,** *83,* 31–40.
45. Picot, A.; Lacroix, C. *J. Food Sci*. **2003,** *68,* 2693–2700.
46. Palchesko, R. N.; Lathrop, K. L.; Funderburgh, J. L.; Feinberg, A. W. *Sci. Rep*. **2015,** *5,* 7955.
47. Giraud-Guille, M. M.; Helary, C.; Vigier, S.; Nassif, N. *Soft Matter*. **2010,** *6,* 4963–4967.
48. Bernal, J. D. *Proc. Phys. Soc. Sect. A* **1949,** *62,* 537.
49. Oparin, A. I. *The Origin of Life on the Earth*, 3rd ed.; Academic Press: New York, 1957.
50. Brack, A. Clay Minerals and the Origin of Life. In *Handbook of Clay Science*; Bergaya, F., Theng, B., Lagaly, K. G. G., Eds.; Elsevier: Amsterdam, 2006; Vol 1, pp 379–391.

51. Fukushima, Y.; Okada, A.; Kawasumi, M.; Kurauchi, T.; Kamigaito, O. *Clay Miner.* **1988**, *23*, 27–34.

52. Kojima, Y.; Usuki, A.; Kawasumi, M.; Okada, A.; Fukushima, Y.; Kurauchi, T.; Kamigaito, O. *J. Mater. Res.* **1993**, *8*, 1185–1189.

53. Fukushima, Y.; Tani, M. *J. Chem. Soc. Chem. Commun.* **1995**, *2*, 241–242.

54. Ray, S. S.; Bousmina, M. *Prog. Mater. Sci.* **2005**, *50*, 962–1079.

55. Ruiz-Hitzky, E.; Van Meerbeek, A. Clay Mineral and Organoclay: Polymer Nanocomposites. In *Handbook of Clay Science*; Bergaya, F., Theng, B. K. G., Lagaly, G., Eds.; Elsevier: Amsterdam, 2006; Vol 1, pp 583–621.

56. Darder, M.; Colilla, M.; Ruiz Hitzky, E. *Chem. Mater.* **2003**, *15*, 3774–3780.

57. Darder, M.; Colilla, M.; Ruiz Hitzky, E. *Appl. Clay Sci.* **2005**, *28*, 199–208.

58. Olmo, N.; Lizarbe, M. A.; Gavilanes, J. G. *Biomaterials* **1987**, *8*, 67–69.

59. Giménez, B.; Gómez Guillén, M. C.; López Caballero, M. E.; Gómez Estaca, J.; Montero, P. *Food Hydrocoll.* **2012**, *27*, 475–486.

60. Darder, M.; Lopez Blanco, M.; Aranda, P.; Aznar, A. J.; Bravo, J.; Ruiz Hitzky, E. *Chem. Mater.* **2006**, *18*, 1602–1610.

61. Ruiz Hitzky, E.; Darder, M.; Aranda, P.; del Burgo, M. A. M.; del Real, G. *Adv. Mater.* **2009**, *21*, 4167–4171.

62. Fernandes, F. M.; Ruiz, A. I.; Darder, M.; Aranda, P.; Ruiz Hitzky, E. *J. Nanosci. Nanotechnol.* **2009**, *9*, 221–229.

63. Wicklein, B.; Darder, M.; Aranda, P.; Ruiz Hitzky, E. *Langmuir* **2010**, *26*, 5217–5225.

64. Alcântara, A. C. S.; Darder, M.; Aranda, P.; Ruiz-Hitzky, E. *Eur. J. Inorg. Chem.* **2012**, 5216–5224.

65. Fernandes, F. M.; Manjubala, I.; Ruiz Hitzky, E. *Phys. Chem. Chem. Phys.* **2011**, *13*, 4901–4910.

66. An, N.; Zhou, C. H.; Zhuang, X. Y.; Tong, D. S.; Yu, W. H. *Appl. Clay Sci.* **2015**, *114*, 283–296.

67. Nalawade, P.; Aware, B.; Kadam, V. J.; Hirlekar, R. S. *Ind. Res.* **2009**, *68*, 267–272.

68. Li, Y.; Liu, D.; Ai, H.; Chang, Q.; Liu, D.; Xia, Y.; Liu, S.; Peng, N.; Xi, Z.; Yang, X. *Nanotechnology* **2010**, *21*, 105101.

69. Darder, M.; Lopez Blanco, M.; Aranda, P.; Leroux, F.; Ruiz Hitzky, E. *Chem. Mater.* **2005**, *17*, 1969–1977.

70. Ruiz Hitzky, E.; Darder, M.; Aranda, P. *J. Mater. Chem.* **2005**, *15*, 3650–3662.

71. Forano, C.; Vial, S.; Mousty, C. *Curr. Nanosci.* **2006**, *2*, 283–294.

72. Choy, J.; Choi, S.; Oh, J.; Park, T. *Appl. Clay Sci.* **2007**, *36*, 122–132.

73. Alcantara, A.; Aranda, P.; Darder, M.; Ruiz Hitzky, E. *J. Mater. Chem.* **2010**, *20*, 9495–9504.

74. Ribeiro, L. N. M.; Alcantara, A. C. S.; Darder, M.; Aranda, P.; Araujo Moreira, F. M.; Ruiz Hitzky, E. *Int. J. Pharm.* **2014**, *463*, 1–9.

75. Pu, Z.; Mark, J. E.; Jethmalani, J. M.; Ford, W. T. *Polym. Bull.* **1996**, *37*, 545–551.

76. Watzke, H. J.; Dieschbourg, C. *Adv. Colloid Interface Sci.* **1994**, *50*, 1–14.

77. Maeda, S.; Armes, S. P. *J. Mater. Chem.* **1994**, *4*, 935–942.

78. Johnson, P.; Whateley, T. L. *J. Colloid Interface Sci.* **1971**, *37*, 557–563.

79. Avnir, D.; Braun, S.; Lev, O.; Ottolenghit, M. *Chem. Mater.* **1994**, *6*, 1605–1614.

80. Nassif, N.; Bouvet, O.; Noelle Rager, M.; Roux, C.; Coradin, T.; Livage, J. *Nat. Mater.* **2002**, *1*, 42–44.

81. Nieto, A.; Areva, S.; Wilson, T.; Viitala, R.; Vallet-Regi, M. *Acta Biomater.* **2009**, *5*, 3478–3487.

82. Fukushima, Y.; Okamura, K.; Imai, K.; Motai, H. *Biotechnol. Bioeng.* **1988**, *32*, 584–594.

83. Xu, S. W.; Jiang, Z. Y.; Lu, Y.; Wu, H.; Yuan, W. K. *Ind. Eng. Chem. Res.* **2006**, *45*, 511–517.

84. Hsu, A. F.; Wu, E.; Folglia, T. A.; Piazza, G. J. J. *Food Biochem.* **1999**, *24*, 21–32.

85. Konsoula, Z.; Liakopoulou Kyriakides, M. *Process Biochem.* **2006**, *41*, 343–349.

86. Khani, Z.; Jolivalt, C.; Cretin, M.; Tingry, S.; Innocent, C. *Biotechnol. Lett.* **2006**, *28*, 1779–1786.

87. Moridi, Z.; Mottaghitalab, V.; Haghi, A. K. *Cell. Chem. Technol.* **2011**, *45*, 549–563.

88. Haider, S.; Park, S. Y.; Saeed, K.; Farmer, B. L. *Sens. Actuators B: Chem.* **2007**, *124*, 517–528.

89. MacDonald, R. A.; Laurenzi, B. F.; Viswanathan, G.; Ajayan, P. M.; Stegemann, J. P. *J. Biomed. Mater. Res. A* **2005**, *74A*, 489–496.

90. Ajayan, P. M.; Stephan, O.; Colliex, C.; Trauth, D. *Science* **1994**, *265*, 1212–1214.

91. Hammel, E.; Tang, X.; Trampert, M.; Schmitt, T.; Mauthner, K.; Eder, A.; Potschke, P. *Carbon N.Y.* **2004**, *42*, 1153–1158.

92. Coleman, J. N.; Khan, U.; Blau, W. J.; Gun'ko, Y. K. *Carbon N.Y.* **2006**, *44*, 1624–1652.

93. Grossiord, N.; Loos, J.; Regev, O.; Koning, C. E. *Chem. Mater.* **2006**, *18*, 1089–1099.

94. Bauhofer, W.; Kovacs, J. Z. *Compos. Sci. Technol.* **2009**, *69*, 1486–1498.

95. Tasis, D.; Tagmatarchis, N.; Bianco, A.; Prato, M. *Chem. Rev.* **2006**, *106*, 1105–1136.

96. Jacobs, C. B.; Peairs, M. J.; Venton, B. *J. Anal. Chim. Acta* **2010**, *662*, 105–127.

97. Szleifer, I.; Yerushalmi Rozen, R. *Polymer* **2005**, *46*, 7803–7818.

98. Yerushalmi Rozen, R.; Szleifer, I. *Soft Matter* **2006**, *2*, 24–28.

99. Wallace, G. G.; Chen, J.; Li, D.; Moulton, S. E.; Razal, J. M. *J. Mater. Chem.* **2010**, *20*, 3553–3562.

100. Liu, J.; Han, L.; Wang, T.; Hong, W.; Liu, Y.; Wang, E. *Chem. Asian J.* **2012**, *7*, 2824–2829.

101. Nicolosi, V.; Chhowalla, M.; Kanatzidis, M. G.; Strano, M. S.; Coleman, J. N. *Science* **2013**, *340*, 1226419–1226419.

102. Filip, J.; Tkac, J. *Electrochim. Acta* **2014**, *136*, 340–354.

103. Pumera, M. *Chem. Soc. Rev.* **2010**, *39*, 4146–4157.

104. Brownson, D. A. C.; Banks, C. E. *Analyst* **2010**, *135*, 2768–2778.

105. Halma, M.; Mousty, C.; Forano, C.; Sancelme, M.; Besse Hoggan, P.; Prevot, V. *Colloids Surf. B: Biointerfaces* **2015**, *126*, 344–350.

106. Schiffer, L.; Anderko, S.; Hobler, A.; Hannemann, F.; Kagawa, N.; Bernhardt, R. *Microb. Cell Fact* **2015**, *14*, 1–12.

107. Ito, Y.; Fukusaki, E. *J. Mol. Catal. B: Enzym.* **2004**, *28*, 155–166.

108. Galandova, J.; Ziyatdinova, G.; Labuda, J. *Anal. Sci.* **2008**, *24*, 711–716.

109. Kaushik, A.; Solanki, P. R.; Ansari, A. a.; Malhotra, B. D.; Ahmad, S. *Biochem. Eng. J.* **2009**, *46*, 132–140.

110. Baum, D. A.; Silverman, S. K. *Cell. Mol. Life Sci.* **2008**, *65*, 2156–2174.

111. Liu, F.; Zhang, J.; Chen, R.; Chen, L.; Deng, L. *Chem. Biodivers.* **2011**, *8*, 311–316.

112. Li, T.; Wang, E.; Dong, S. *Chem. Commun.* **2008**, 5520–5522.

113. Schlosser, K.; Li, Y. *Chem. Biol.* **2009**, *16*, 311–322.

114. Wei, W. L.; Zhu, H. Y.; Zhao, C. L.; Huang, M. Y.; Jiang, Y. Y. *React. Funct. Polym.* **2004,** *59,* 33–39.

115. Desimone, M. F.; Helary, C.; Rietveld, I. B.; Bataille, I.; Mosser, G.; Giraud Guille, M.M.; Livage, J.; Coradin, T. *Acta Biomater.* **2010,** *6,* 3998–4004.

116. Lin, C. C.; Chu, Y. M.; Chang, H. C. *Sens. Actuators, B: Chem.* **2013,** *187,* 533–539.

117. Kaushik, A.; Solanki, P. R.; Ansari, A. a.; Ahmad, S.; Malhotra, B. D. *Electrochem. Commun.* **2008,** *10,* 1364–1368.

118. Shan, D.; Wang, S.; Xue, H.; Cosnier, S. *Electrochem. Commun.* **2007,** *9,* 529–534.

119. Xu, H.; Dai, H.; Chen, G. *Talanta* **2010,** *81,* 334–338.

120. Archelas, A.; Demirjian, D. C.; Furstoss, R.; Griengl, H.; Jaeger, K. E.; Moris Varas, F.; Ohrlein, R.; Reetz, M. T.; Reymond, J. L.; Schmidt, M.; Servi, S.; Shah, P. C.; Tischer, W.; Wedekind, F. In *Biocatalysis: From Discovery to Application*; Fessner, W. D., de Meijere, A., Kessler, H., Lay, S. V., Thien, J., Vogtle, F., Houk, K. N., Lehn, J. M., Schreiber, S. L., Trost, B. M., Yamamoto, H., Eds.; Springer: Berlin, 1999.

121. Hasan, F.; Shah, A. A.; Hameed, A. *Enzyme Microb. Technol.* **2006,** *39,* 235–251.

122. Chen, X.; Cheng, G.; Dong, S. *Analyst* **2001,** *126,* 1728–1732.

123. Tan, X.; Li, M.; Cai, P.; Luo, L.; Zou, X. *Anal. Biochem.* **2005,** *337,* 111–120.

124. Shi, Q.; Han, E.; Shan, D.; Yao, W.; Xue, H. *Bioprocess Biosyst. Eng.* **2008,** *31,* 519–526.

125. Sheldon, R. A. *Adv. Synth. Catal.* **2007,** *349,* 1289–1307.

126. Heichal Segal, O.; Rappoport, S.; Braun, S. *Biotechnology* **1995,** *13,* 798–800.

127. Murua, A.; Portero, A.; Orive, G.; Hernandez, R. M.; de Castro, M.; Pedraz, J. L. *J. Control. Release* **2008,** *132,* 76–83.

128. Dicosmo, F.; Misawa, M. *Biotechnol. Adv.* **1995,** *13,* 425–453.

129. Ramachandra Rao, S.; Ravishankar, G. A. *Biotechnol. Adv.* **2002,** *20,* 101–153.

130. Leonard, A.; Dandoy, P.; Danloy, E.; Leroux, G.; Meunier, C. F.; Rooke, J. C.; Su, B. L. *Chem. Soc. Rev.* **2011,** *40,* 860–885.

131. Oczkowski, T.; Zwierkowska, E.; Bartkowiak, S. *Bioelectrochemistry* **2007,** *70,* 192–197.

132. Pilon Smits, E. *Annu. Rev. Plant Biol.* **2005,** *56,* 15–39.

133. Vila, M.; Pascal Larber, S.; Rathahao, E.; Debrauwer, L.; Canlet, C.; Laurent, F. *Environ. Sci. Technol.* **2005,** *39,* 663–672.

134. Amaro, H. M.; Macedo, A. C.; Malcata, F. X. *Energy* **2012,** *44,* 158–166.

135. Birolli, W. G.; Ferreira, I. M.; Alvarenga, N.; Santos, D. D.; de Matos, I. L.; Comasseto, J. V.; Porto, A. L. M. *Biotechnol. Adv.* **2015,** *33,* 481–510.

136. Shitanda, I.; Takamatsu, S.; Watanabe, K.; Itagaki, M. *Electrochim. Acta* **2009,** *54,* 4933–4936.

137. Mallick, N. *BioMetals* **2002,** *15,* 377–390.

138. Shimoda, K.; Hamada, H. *Environ. Health Insights* **2009,** *3,* 89–94.

139. Sanchez, C. *Biotechnol. Adv.* **2009,** *27,* 185–194.

140. Pointing, S. B. *Appl. Microbiol. Biotechnol.* **2001,** *57,* 20–33.

141. Baronian, K. H. R. *Biosens. Bioelectron.* **2004,** *19,* 953–962.

142. Fukuda, H.; Hama, S.; Tamalampudi, S.; Noda, H. *Trends Biotechnol.* **2008,** *26,* 668–673.

143. Schrewe, M.; Julsing, M. K.; Buhler, B.; Schmid, A. *Chem. Soc. Rev.* **2013,** *42,* 6346–6377.

144. Logan, B. E.; Call, D.; Cheng, S.; Hamelers, H. V. M.; Sleutels, T. H. J. A.; Jeremiasse, A. W.; Rozendal, R. A. *Environ. Sci. Technol.* **2008**, *42*, 8630–8640.

145. Rabaey, K.; Verstraete, W. *Trends Biotechnol.* **2005**, *23*, 291–298.

146. D'Souza, S. F. *Biosens. Bioelectron.* **2001**, *16*, 337–353.

147. Belkin, S. *Curr. Opin. Microbiol.* **2003**, *6*, 206–212.

148. Kulkarni, M.; Chaudhari, A. *J. Environ. Manage.* **2007**, *85*, 496–512.

149. Valls, M.; de Lorenzo, V. *FEMS Microbiol. Rev.* **2002**, *26*, 327–338.

150. Khan, R.; Bhawana, P.; Fulekar, M. H. *Rev. Environ. Sci. Biotechnol.* **2012**, *12*, 75–97.

151. Peng, R. H.; Xiong, A. S.; Xue, Y.; Fu, X. Y.; Gao, F.; Zhao, W.; Tian, Y. S.; Yao, Q. H. *FEMS Microbiol. Rev.* **2008**, *32*, 927–955.

152. Ohtsubo, Y.; Kudo, T.; Tsuda, M.; Nagata, Y. *Appl. Microbiol. Biotechnol.* **2004**, *65*, 250–258.

153. Michelini, E.; Roda, A. *Anal. Bioanal. Chem.* **2012**, *402*, 1785–1797.

154. Bruna, F.; Pereira, M. G.; Polizeli, M. de L. T. M.; Valim, J. B. *ACS Appl. Mater. Interfaces* **2015**, *7*, 18832–18842.

155. Benaissi, K.; Helaine, V.; Prevot, V.; Forano, C. *Adv. Synth. Catal.* **2011**, *353*, 1497–1509.

156. Carrasco, M.; Rad, J. C.; Gonzalez Carcedo, S. *Bioresour. Technol.* **1995**, *51*, 175–181.

157. Sanjay, G.; Sugunan, S. *Catal. Commun.* **2006**, *7*, 1005–1011.

158. Huang, J.; Liu, Y.; Wang, X. J. *Mol. Catal. B: Enzym.* **2009**, *57*, 10–15.

159. Tomar, M.; Prabhu, K. A. *Enzyme Microb. Technol.* **1985**, *7*, 454–458.

160. Sanjay, G.; Sugunan, S. *Food Chem.* **2006**, *94*, 573–579.

161. Chang, M. Y.; Juang, R. S. *Enzyme Microb. Technol.* **2005**, *36*, 75–82.

162. Chang, M. Y.; Kao, H. C.; Juang, R. S. *Int. J. Biol. Macromol.* **2008**, *43*, 48–53.

163. Coradin, T.; Nassif, N.; Livage, J. *Appl. Microbiol. Biotechnol.* **2003**, *61*, 429–434.

164. Coradin, T.; Mercey, E.; Lisnard, L.; Livage, J. *Chem. Commun.* **2001**, 2496–2497.

165. Schuleit, M.; Luisi, P. L. *Biotechnol. Bioeng.* **2001**, *72*, 249–253.

166. Yilmaz, E.; Sezgin, M.; Yilmaz, M. *J. Mol. Catal. B: Enzym.* **2010**, *62*, 162–168.

167. Liria, C. W.; Ungaro, V. A.; Fernandes, R. M.; Costa, N. J. S.; Marana, S. R.; Rossi, L. M.; Machini, M. T. *J. Nanopart. Res.* **2014**, *16*, 2612.

168. Gonzalez Dominguez, E.; Comesana Hermo, M.; Marino Fernandez, R.; Rodriguez Gonzalez, B.; Arenal, R.; Salgueirino, V.; Moldes, D.; Othman, A. M.; Perez Lorenzo, M.; Correa Duarte, M. A. *ChemCatChem* **2016**, *8*, 1264–1268.

169. Liao, C. A.; Wu, Q.; Wei, Q. C.; Wang, Q. G. *Chem. Eur. J.* **2015**, *21*, 12620–12626.

170. Miao, Y.; Tan, S. N. *Anal. Chim. Acta* **2001**, *437*, 87–93.

171. Li, W.; Yuan, R.; Chai, Y.; Zhou, L.; Chen, S.; Li, N. *J. Biochem. Biophys. Methods* **2008**, *70*, 830–837.

172. Jia, N.; Zhou, Q.; Liu, L.; Yan, M.; Jiang, Z. *J. Electroanal. Chem.* **2005**, *580*, 213–221.

173. Fernandes, F. M.; Ruiz-Hitzky, E. *Carbon N.Y.* **2014**, *72*, 296–303.

174. Solanki, P. R.; Kaushik, A.; Ansari, A. A.; Tiwari, A.; Malhotra, B. D. *Sens. Actuators, B: Chem.* **2009**, *137*, 727–735.

175. Shan, D.; Cosnier, S.; Mousty, C. *Anal. Chem.* **2003**, *75*, 3872–3879.

176. Mousty, C. *Appl. Clay Sci.* **2004**, *27*, 159–177.

177. Liu, L. M.; Jiang, L. P.; Liu, F.; Lu, G. Y.; Abdel Halim, E. S.; Zhu, J. *Anal. Methods* **2013**, *5*, 3565–3571.

178. Ding, S. N.; Shan, D.; Xue, H. G.; Zhu, D. B.; Cosnier, S. *Anal. Sci.* **2009,** *25,* 1421–1425.

179. Ding, S. N.; Shan, D.; Zhang, T.; Dou, Y. Z. *J. Electroanal. Chem.* **2011,** *659,* 1–5.

180. Han, E.; Shan, D.; Xue, H.; Cosnier, S. *Biomacromolecules* **2007,** *8,* 971–975.

181. Lopez, M. S. P.; Leroux, F.; Mousty, C. *Sens. Actuators, B: Chem.* **2010,** *150,* 36–42.

182. Le Ouay, B.; Coradin, T.; Laberty Robert, C. *J. Mater. Chem. B* **2013,** *1,* 606.

183. Kuncova, G.; Podrazky, O.; Ripp, S.; Trogl, J.; Sayler, G. S.; Demnerova, K.; Vankova, R. *J. Sol Gel Sci. Technol.* **2004,** *31,* 335–342.

184. Ramachandran, S.; Coradin, T.; Jain, P. K.; Verma, S. K. *Silicon* **2010,** *1,* 215–223.

185. Perullini, M.; Jobbagy, M.; Mouso, N.; Forchiassin, F.; Bilmes, S. A. *J. Mater. Chem.* **2010,** *20,* 6479–6483.

186. Tong, H. W.; Mutlu, B. R.; Wackett, L. P.; Aksan, A. *Mater. Lett.* **2013,** *111,* 234–237.

187. Perullini, M.; Ferro, Y.; Durrieu, C.; Jobbagy, M.; Bilmes, S. A. *J. Biotechnol.* **2014,** *179,* 65–70.

188. Mandal, S.; Patil, V. S.; Mayadevi, S. *Microporous Mesoporous Mater.* **2012,** *158,* 241–246.

189. Alekseeva, T.; Prevot, V.; Sancelme, M.; Forano, C.; Besse-Hoggan, P. *J. Hazard. Mater.* **2011,** *191,* 126–135.

190. Hsu, A. F.; Jones, K.; Foglia, T. a.; Marmer, W. N. *Biotechnol. Appl. Biochem.* **2002,** *36,* 181–186.

191. Mayer, A. M.; Staples, R. C. *Phytochemistry* **2002,** *60,* 551–565.

192. Torres, E.; Bustos-Jaimes, I.; Le Borgne, S. *Appl. Catal. B: Environ.* **2003,** *46,* 1–15.

193. Trivedi, U. J.; Bassi, A. S.; Zhu, J. *Can. J. Chem. Eng.* **2006,** *84,* 239–247.

194. Verstraete, W.; Rabaey, K.; Logan, B. E.; Hamelers, B.; Rozendal, R.; Schroder, U.; Keller, J.; Freguia, S.; Aelterman, P.; Verstraete, W.; Rabaey, K. *Environ. Sci. Technol.* **2006,** *40,* 5181–5192.

195. Velasquez Orta, S. B.; Curtis, T. P.; Logan, B. E. *Biotechnol. Bioeng.* **2009,** *103,* 1068–1076.

196. Yuan, Y.; Chen, Q.; Zhou, S.; Zhuang, L.; Hu, P. *J. Hazard. Mater.* **2011,** *187,* 591–595.

197. Posten, C.; Schaub, G. *J. Biotechnol.* **2009,** *142,* 64–69.

198. Ariga, K.; Ji, Q.; Mori, T.; Naito, M.; Yamauchi, Y.; Abe, H.; Hill, J. P. *Chem. Soc. Rev.* **2013,** *42,* 6322–6345.

199. Ma, J.; Sahai, Y. *Carbohydr. Polym.* **2013,** *92,* 955–975.

200. Song, T. S.; Jin, Y.; Bao, J.; Kang, D.; Xie, J. *J. Hazard. Mater.* **2016,** *317,* 73–80.

201. Zou, L.; Qiao, Y.; Wu, X. S.; Li, C. M. *J. Power Sources* **2016,** *328,* 143–150.

202. Desmet, J.; Meunier, C.; Danloy, E.; Duprez, M. E.; Lox, F.; Thomas, D.; Hantson, A. L.; Crine, M.; Toye, D.; Rooke, J.; Su, B. L. *J. Colloid Interface Sci.* **2015,** *448,* 79–87.

203. Srirangan, K.; Pyne, M. E.; Perry Chou, C. *Bioresour. Technol.* **2011,** *102,* 8589–8604.

204. Cooney, M. J.; Svoboda, V.; Lau, C.; Martin, G.; Minteer, S. D. *Energy Environ. Sci.* **2008,** *1,* 320–337.

205. Deng, L.; Shang, L.; Wen, D.; Zhai, J.; Dong, S. *Biosens. Bioelectron.* **2010,** *26,* 70–73.

206. Wang, Z. G.; Ke, B. B.; Xu, Z. K. *Biotechnol. Bioeng.* **2006,** *97,* 708–720.

207. Derby, B. *J. Mater. Chem.* **2008,** *18,* 5717–5721.

CHAPTER 10

A Study on Magnetic Metal Carbon Mesocomposites Green Synthesis Peculiarities with Point of Chemical Mesoscopics View

V. I. KODOLOV[1,2*], V. V. KODOLOVA–CHUKHONTZEVA[1,3], N. S. TEREBOVA[1,4], and I. N. SHABANOVA[1,4]

[1]*Basic Research, High Educational Centre of Chemical Physics and Mesoscopics, UD of RAS, Izhevsk, Russian Federation*

[2]*M.T. Kalashnikov Izhevsk State Technical University, Izhevsk, Russia*

[3]*Institute of Macromolecular Compounds, Russian Academy of Sciences, Saint Petersburg, Russia*

[4]*Udmurt Federal Research Centre, Ural Division, RAS, Izhevsk, Russia*

*Corresponding author. E-mail: vkodol.av@mail.ru

ABSTRACT

The mechanism of mesoparticle modification reactions is considered with the application of such notions as charges quantization, phase coherence, interference, and annihilation. On the basis of theoretical mesoscopics ideas, the formation of covalent bonds because of the interference of negative charges quants in modification reactions is discussed. The hypothesis about possibility of annihilation at the interaction of positive and negative charges quants in redox processes is presented. The magnetic metal carbon mesoscopic composites synthesis (e.g., initial metal carbon mesocomposites) is realized by mechanochemical method of the grinding of metal oxides microscopic particles with polyvinyl alcohol macromolecules.

Then, in the result, the copper or nickel carbon mesocomposites, which have the following atomic magnetic moments: for copper—1.3 μ_B, for nickel—1.8 μ_B, are obtained. The investigations are carried out on the analysis examples of processes of copper and nickel carbon mesoparticles modification by the compounds containing p, d elements. In the middle, substances such as polyethylene polyamine, ammonium iodide, ammonium polyphosphate (APPh), silica (SiO_2), aluminium oxide, iron oxide, nickel oxide, and copper oxide are used. It is noted that the redox processes are accompanied by the metal atomic magnetic moments growth which is explained by the electron shift on high energetic levels because of the annihilation phenomenon. The hypothesis concerning the passing of two phenomena (annihilation and interference) at redox processes is proposed.

10.1   INTRODUCTION

This chapter is dedicated to the continuation of the copper and nickel carbon mesopaticles modification process investigations. The modification processes are realized by mechanochemical method of the grinding of metal carbon mesoscopic composites (Cu-C MC or Ni-C MC) with such compounds as polyvinyl alcohol (PVA), polyethylene polyamine (PEPA), ammonium iodide (AmI), ammonium polyphosphate (APPh), silica (SiO_2), aluminium oxide (Al_2O_3), iron oxide, nickel oxide, and copper oxide. However, in this chapter, the investigation results are discussed for systems such as "Cu-C MC – APPh (or silica)", "Ni –C MC – APPh (or silica)", and also for the system "Ni-C MC – Al_2O_3". For the investigations of these systems, X-ray photoelectron spectroscopy, electron paramagnetic resonance, and transition electron microscopy with electron diffraction are used.

10.2   SHORT THEORY

The process occurs at the charges quantization with the certain phase coherence and then with the chemical bond formation because of the interference as well as in the redox processes where possible annihilation takes place.[1, 2] If chemical reactions are realized without the changes in the oxidation state of atoms, then the negative charges quants quantization and the interference are carried out. However, the most reactions flow with the changes in the oxidation state elements and then according to known schemes of reduction–oxidation processes. Therefore, it is necessary to take into consideration of positive charge quants. The interaction of positive charge quants with the negative charge quants, the annihilation phenomenon with the electromagnetic radiation, or/and the direct electromagnetic field are possible. Also, the interaction of positive charges with the formation of "dark hole" must not be excluded (Fig. 10.1). In this case, the explosion with diffusion of most quantity of energy into surroundings is possible.

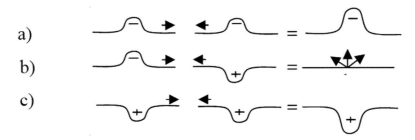

FIGURE 10.1   The scheme of charge quants interaction with appearance of interference (a), annihilation (b), and the formation of "dark hole" (c).

The phenomena of charge quantization, interference, and annihilation are considered on the examples of metal carbon mesoparticle interactions with reagents containing p and d elements. For these investigations, the basic method of research is X-ray photoelectron spectroscopy.

10.3   RESULTS AND DISCUSSION

The production of metal carbon mesoscopic composites is carried out by using of mechanochemical interaction between microscopic particles of metal oxides and macromolecules of polymers at the active medium presence.[3] For mesoscopic composites production, the sign variable loadings are applied. These loadings are appeared at the grinding with pressing.

At the common grinding of copper oxide particles with polyvinyl alcohol (PVA) particles (or concentrated water solution), the metallic phase clusters fall between macromolecules of polyvinyl alcohol or, in other words, into reservoir (according to mesoscopic notions), in which banks are PVA macromolecules. The metal (e.g., copper) within cluster has the positive charge. Therefore, the negative charge quants are directed to positive charged atom. In our case, the negative charged quants from polyvinyl alcohol acetate and hydroxyl groups are transferred to copper positive charge quants. As a result, the annihilation with the electromagnetic direct field formation takes place. In this process, the acetic acid and water are formed, and also the bank's structures are changed: the polyacetylene and carbine fragments appear. There are unpaired electrons on joints of these fragments. The process of pair electron division and the shift of electrons on the high atomic levels for metal are explained by the annihilation origin. In this case, the metal atomic magnetic moment growth is observed in the dependence on the electrons number, which participates in the redox process.

The hypothesis about possibility of annihilation at the interaction of positive and negative charges quants in redox processes is confirmed by the examples of processes of copper and nickel carbon mesocomposites modification with the application of substances such as polyethylene polyamine, ammonium iodide, ammonium polyphosphate (APP), silica (SiO_2), aluminium oxide, iron oxide, nickel oxide, and copper oxide.[3–5] When polyethylene polyamine and ammonium iodide are applied, the connection reactions take place. At the interactions of polyethylene polyamine with mesoparticles, the C=N bond formation is explained by the interference of negative charge quants. When the mesoparticles modification reactions using APP, SiO_2, and metal oxides are carried out, the redox processes are realized. In these cases, the modifiers reduction reactions take place.[6–8] The structures of metal carbon mesoscopic composites with active carbon shells are defined by means of the complex of methods including X-ray photoelectron spectroscopy, transition electron microscopy with high permission, electron micro diffraction, and EPR spectroscopy.

In correspondent reactions, the element reduction for reagents and nickel or copper atomic magnetic moments growth in mesoparticles takes place. In Table 10.1, the examples of change in metal atomic magnetic moment for mesoparticles modified by APPh or silica after the mechanochemical modification processes are given.

TABLE 10.1 The Values of Copper (Nickel) Atomic Magnetic Moments in the Interaction Products for Systems:

Cu-C MC – APPh (or SiO_2) and Ni-C MC – APPh (or SiO_2).

Systems Cu C NC – substances	μ_{cu}	Systems Ni C NC – substance	μ_{Ni}
Cu C NC – silica	3.0	Ni C NC – silica	4.0
Cu C NC ΛPPh	2.0	Ni C NC – ΛPPh	3.0
Cu C NC – APPh, relation 1:0.5	4.2		

The presence of unpaired electrons on msoparticles carbon shells in above systems is determined by means of electron paramagnetic resonance (EPR) (Table 10.2).

TABLE 10.2 The Unpaired Electron Values (from EPR Spectra) for Systems "Cu-C NC – silica" and "Cu/C NC – APPh" (relation 1:1) in Comparison with Mesoparticle Cu/C NC.

Substance	Quantity of unpaired electrons, spin/g
Cu/C nanocomposite	1.2×10^{17}
system Cu/C NC – SiO_2	3.4×10^{19}
system Cu/C NC – APPh	2.8×10^{18}

The metal atomic magnetic moment growth proceeds owing to the redox processes with above chemical compounds. In Refs. [4, 5], it has been shown that the reduction reactions of phosphorus and silicon from correspondent substances at the interaction on the interphase boundary with mesoparticles are realized.[4, 5]

The relations of mesoparticles to above oxides are changed from 1:1 to 1:0.2 depending on the obtained qualitative spectra, for example, the relations of 1:1 and 1:0.5 for system "Ni/C NC – Al_2O_3" lead to full mask of mesoparticle. Therefore, the quantity of aluminium oxide is decreased to the relation 1:0.2. In accordance with Al3s spectra, aluminum is completely reduced during the modification process, and nickel atomic magnetic moment is increased to $4.8\mu_B$ (Table 10.3).

TABLE 10.3 Parameters of Multiple Splintering Me3s Spectra in System Ni/C + Al_2O_3.

Sample	I_2/I_1 (Ni)	Δ_{Ni}, eV	$\mu_{Ni}\mu_B$
Ni C MC	0.3	3.0	1.8
Ni C MC + Al_2O_3 = 1:0.2	0.9	1.0	4.8

Here, I_2/I_1 is the relation of multiple splintering lines maximums intensities and

Δ is the energetic distance between the multiple splintering maximums in Me3s spectra.

In this case, the reduction process is related to not only aluminum oxide, but also to nickel oxide from metal cluster of mesoparticle (Ni C MC). Therefore, the reduction processes are stipulated by the electron transport from carbon shell of mesoparticle in direction to Al^{+3} and Ni^{+2} of atoms in correspondent oxides.

Exceptional properties of modified metal carbon nanostructures with magnetic characteristics lead to the properties improvement of nanostructured polymeric coatings. For example, the introduction of 0.008% Cu C MC into the melamine–formaldehyde resins stimulates the polarization growth in two-times (on the AFM data). Similar results can be received at the combination of phase coherency and interference of charge quants during the preparation process of modified polymeric materials. The decreasing of nanostructures activity is possible when the modification is carried out with the ultrasound processing or the violation of phase coherency takes place as the nanostructure quantity changes.

10.4   CONCLUSIONS

The present investigation has a fundamental character. It is based on the ideas concerning the change of metal carbon mesoscopic composites reactivity. The investigations are dedicated to the mechanochemical redox processes in which the electron transport from mesoscopic composite cluster to carbon shell takes place. In this case, the electron delocalization is found. For the first time, the metal carbon mesoscopic composite modification by mechanochemical process with the using of active substances including also bioactive systems is possible. The activity of metal carbon mesoscopic composites is caused by the structure and composition of correspondent composites, which contain the delocalized electrons and double bonds on the surface of carbon shell.

Thus, at the mechanic chemical reduction/oxidation synthesis the changes of element oxidation states as well as the increasing of metal atomic moment for cluster can appear. At the same time, the modifiers elements and functional groups are discovered in carbon shell of mesoscopic composites modified.

The creation of reactive mesoscopic materials with regulated magnetic characteristics which can find the application as modifiers of materials properties, catalysts for different processes, effective inhibitors of corrosion, sorbents, stimulators of plant growth is very topical. These facts open new era for further investigations and development of metal carbon mesoscopic composites application fields.

KEYWORDS

- **charges quantization**
- **phase coherency**
- **interference**
- **annihilation**
- **mesoparticles**
- **metal carbon mesocomposites**
- **redox processes**
- **metal atomic magnetic moment**

REFERENCES

1. Kodolov, V. I.; Trineeva, V. V. New Scientific Trend – Chemical Mesoscopics. *Chem. Phys. Mesoscopics*. **2017,** *19*(3), 454–465.
2. Yavorskiy, B. M.; Detlaph, A. A. Directory on Physics. M.: Publ. "Science". 1965, p.848. Kuhling H. Reference Book on Physics. M.: Publ. "World". 1983, p. 519.
3. Rudenberg, K. *Physical Nature of Chemical Bond*. M.: Publ. "World". 1964, p.162.
4. Kodolov, V. I.; Trineeva, V. V.; Kopylova, A. A. et al. Mechanochemical Modification of Metal/Carbon Nanocomposites. *Chem. Phys. Mesoscopics*. **2017,** *19*(4), 569–580.
5. Shabanova, I. N.; Kodolov, V. I.; Terebova, N. S.; Trineeva, V. V. X ray Electron Spectroscopy in Investigations of Metal/Carbon Nanosystems and Nanostructured Materials. M.-Izhevsk: Publ. "Udmurt University". 2012, p. 252.
6. Kodolov, V. I.; Trineeva, V. V.; Terebova, N. S. et al. Changes of Electron Structure and Magnetic Characteristics of Modified Copper/Carbon Nanocomposites. *Chem. Phys. Mesoscopics*. **2018,** *20*(1), 72–79.
7. Kopylova, A.A.; Kodolov, V. V. Investigation of Coper/Carbon Nanocomposite Interaction with Silicium Atoms from Silicon Compounds. *Chem. Phys. Mesoscopics*. **2014,** *16*(4), 556–560.
8. Wang, J. Q.; Wu, W. M.; Feng, D. M. *The Introduction to Electron Spectroscopy (XPS/XAES/UPS)*; National Definite Industry Press: Beijing, 1992; p 640.

CHAPTER 11

The Role of Nanostructure Activity in Selected Green Material Modification

V. I. KODOLOV[1,2], V. V. KODOLOVA–CHUKHONTZEVA[1,3],
YU V. PERSHIN[1], R. V. MUSTAKIMOV[1], G. I. YAKOVLEV[1,2], and
A. YU. BONDAR[2]

[1]*Basic Research, High Educational Centre of Chemical Physics and Mesoscopics, UD of RAS, Izhevsk, Russian Federation*

[2]*M.T. Kalashnikov Izhevsk State Technical University, Izhevsk, Russia*

[3]*Udmurt Federal Research Centre, Ural Division, RAS, Izhevsk, Russia*

Corresponding author. E-mail: vkodol.av@mail.ru; kodol@istu.ru

ABSTRACT

This chapter is dedicated by the discussion concerning to the investigation's results directed on the designation of possible reasons of the nanostructures or metal-carbon mesocomposites influence decrease on the liquid media and polymeric compositions as well as lowering of positive effect on the properties of the polymeric material at the nanostructures introduction within polymeric compositions. The disturbance of phase coherence can be one of the main reasons for the nanostructure activity of the nanostructure activity decrease. The example of changes in mesoparticles' sizes and their distribution on sizes in the dependence on z-potential for the medium, in which the thin dispersed suspension is prepared, are presented. The method of medium polarization estimation on the changes in peak intensities in IR spectra of mesoparticles suspensions is proposed. It is noted that the changes in media polarization character, which is determined on the peak intensity relations in IR spectra, is changed in the dependence on the quantity of mesoparticles in suspension, and also in depending on time of ultrasound cultivation, which is produced for the improvement of mesoparticles' distribution in the suspension volume. These results are explained by the phase coherence disturbance because of energetic interaction of nanostructure field (the first case) or ultrasound field intensity changes (the second case). The decrease in positive effect or the absence of the nanostructured gas concrete or hard concrete modified by carbon nanotubes is possible owing to the polarization decrease and the phase coherence disturbance because of the polar component (water) removal from composition at the drying process.

11.1 INTRODUCTION

The main purpose of nanostructures or mesoparticles (mesocomposites), as materials modifiers, is the improvement of these materials' properties. This purpose can be achieved at the proportional distribution of nano-structures (mesoparticles) in suspension volume and at their coagulation exclusion. That is possible at the presence of the following conditions: the correspondent polarity, the medium definite dielectric constant, nanostructure's low concentration, and optimal ultrasound action on the correspondent suspension. The last action is necessary for the additional decrease in mesoparticle sizes and for the proportional distribution of mesoparticles in medium nature, which also have influence on the result obtained. Therefore, the great scientific interest presents the grounding of conditions for the material modification with nanostructure (mesoparticle) application without big expenses.

Hence, on the basis of theoretical data and experimental results, it is necessary to define the reasons of nanostructure (or mesoscopic system) activity decrease at the material modification with the application of corre-spondent suspensions (or trademarks) of mesoscopic modifiers.

This chapter is dedicated to the investigation of reasons of decrease in positive effect from the several metal carbon mesocomposites or nano-structures at their introduction in polymeric compositions.

11.2 THE INTERACTION OF LIQIUD MEDIA WITH NANOSTRUCTURES (OR MESOSCOPIC PARTICLES)

The purpose of active nanostructures (or mesoscopic systems) is to improve the properties of modified material. For the determination of obtained organized phase part, the Kolmogorov–Avrami eq 11.1[1,2] is applied:

$$W = 1 - \exp(-k\tau^n), \tag{11.1}$$

Where, W—a part of organized phase; k—a parameter, defined the organized phase growth rate; τ—the duration of organized phase growth; and n—fractal dimension.

Proposed analogs of Avrami equations (11.2) are written by means of the following form:

$$1 - \upsilon = k_1 \exp[-a\, k_T \tau^n], \tag{11.2}$$

$$W = 1 - k_1 \exp\left[-(\varepsilon_S/\varepsilon_V)\tau^n\right] = 1 - k_1 \exp[-(\varepsilon_S^0/\varepsilon_V^0 \cdot S/V)\, k_T \tau^n], \tag{11.3}$$

$$W = 1 - k_1 \exp[-\tau^n(I_{rc}/I_c)k_T], \tag{11.4}$$

Where, a—the nanostructure activity; k_T and k_1—coefficients of proportionality with the registered temperature factors; τ—the duration necessary for the process development to organized phase is formed; n—fractal dimension or index of process direction into the organized phase formation; and I_{rc} and I_c—the intensities of vibrations for the medium chemical particles changed bonds when the medium includes mesocomposite and when the mesocomposite is absent within it.

The following equation for the determination of nanostructures influences on medium is proposed in eq 11.1:

$$W = (n/N)\exp[(an\tau^\beta)/T] \tag{11.5}$$

Where, n—the mesoparticle's quantity within medium; N—the quantity of branches; a—the mesoparticle activity; τ—the duration of synthesis process; T—the temperature of organization of structures; and β—index of process direction for the formation of defined nanostructure forms at the creation of correspondent mesocomposite (this index corresponds to fractal dimension).

It is established that the mesoparticles (or nanostructures) size and their distribution on sizes are defined by the solvent polarity in which mesoparticles are found. In this case, the metal (e.g., copper)-carbon mesocomposites studied, and solvents such as ethanol, acetone, and toluene are applied at the Zeta-potential measurement (Table 11.1). According to the obtained results, the mesoparticles' sizes are decreased and their distribution becomes narrower at the increase in medium polarity or at the decrease in Zeta potential, mV (Zetatrac NPA 152-31 A). The analysis of the obtained results shows that the big mesoparticles of copper carbon nanostructures are formed because of coagulation in nonpolar medium (toluene).

TABLE 11.1 Characteristics of Fine Dispersed Suspensions of Copper-Carbon Mesocomposite in Different Media.[3]

No.	Medium	Zeta-potential, mV	Particles size, nm	Range of particles distribution on size in medium, nm
1	Ethanol	-41.76	15.10–34.43	19.33
2	Acetone	-50.74	10.88–18.91	8.03
3	Toluene	-13.87	32.03–80.1	45.07

It is shown in Ref. [3] that the intensities of peaks for single groups (usually polar groups or bonds) in IR spectra are increased under the influence of metal-carbon mesocomposites because of the molar absorption coefficient growth at the self-organization of system "medium – mesocomposite".[3] Therefore, in the proposed eq 11.4, the relation of intensities, which is equivalently to the relation of molar absorption coefficients, is introduced for IR spectra of media structured by mesocomposites in comparison with IR spectra of media in which these mesocomposites are absent.

On the basis of investigation's results, it has been noted that the changes in peak intensities in IR spectra depend on the quantities of mesoparticles introduced in correspondent medium (Fig. 11.1 and Table 11.2) as well as the duration of correspondent suspension ultrasound processing (Fig. 11.2 and Fig. 11.4).

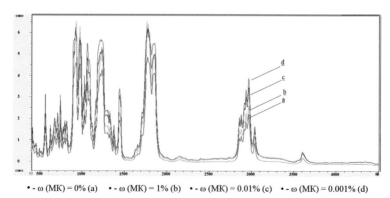

• - ω (MK) = 0% (a)   • - ω (MK) = 1% (b)   • - ω (MK) = 0.01% (c)   • - ω (MK) = 0.001% (d)

FIGURE 11.1 The changes in IR-spectra of fine dispersed suspension of copper-carbon mesocomposite (MC) on the basis of isomethyl tetrahydrophtalic anhydride (ω[MC] = 0.001; 0.01; 1%).[8]

The peak intensity relations for the vibrations of bonds and groups C=O, C–O–C, C–H from IR spectra data lead in Table 11.2.

There are two meanings of intensities relations for symmetric and unsymmetrical vibrations of C=O group, and also for C–H bond vibration at wavenumbers 1450 and 2860–3090 cm^{-1}.

TABLE 11.2 Characteristic Frequencies and Peak Intensity Relations for the Copper-Carbon Mesocomposite Suspension on the Basis of Isomethyl Tetrahydro Phthalic Anhydrate (iMTHPhA) at Different Concentration of Mesocomposite.[8]

No.	v (cm^{-1})	I_1/I_{Θ}	$I_{0.01}/I_{\Theta}$	$I_{0.001}/I_{\Theta}$	Conformity
1	1050	1.235	1.411	1.686	C–O–C st
2	1450	1.179	1.590	1.744	C–H
3	1776	1.458	1.347	1.691	C=O st as
4	1844	1.463	1.412	1.678	C=O st sy
5	2860–3090	1.182	1.545	1.750	C–H

The data presented in Table 11.2 can be reorganized in correspondent graph of dependence of peak intensity relations from the mesoparticles quantity in the suspension (Fig. 11.2).

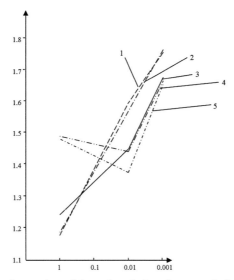

FIGURE 11.2   The changes in peak intensity relations for groups' vibrations in iMTHPhA depending on the nature of group and the quantity of mesoparticles in suspension (1) CH (5); (2) CH (2); (3) COC (1); (4) C=O as (3); (5) C=O sy (4); all numbers are from Table 11.2.

According to Figure 11.2, the approximate linear dependence of the peak intensities increasing for C–H bonds from the copper-carbon meso-composite quantity is observed. The correspondent dependences for C=O bonds (symmetric and unsymmetrical vibrations) and for C–O–C group represent broken lines (Figure 11.2).

The line that reflects the dependence of intensity relations for C=O vibrations has the minimum at the mesocomposite concentration in suspension equaled to 0.01%. It is possible when this fact is explained by the phase coherency disturbance at the increasing mesoparticles' field action. Similar effect is observed for C–O–C line that illustrated the dependence of vibration intensities relation from the mesoparticles quantity. The disturbance of phase coherency leads to the decrease in medium polarization. Therefore, let us consider the possible mechanism of medium polarization under the mesocomposite influence. The mechanism of polarization of suspension obtained can be described by the scheme on Figure 11.3. According to this scheme, the flow of negative charge quants gives the action on the vibration of groups and bonds in the anhydride ring field, especially, in the vibration region of such groups and bonds as C–O–C, C=O, and C–H.

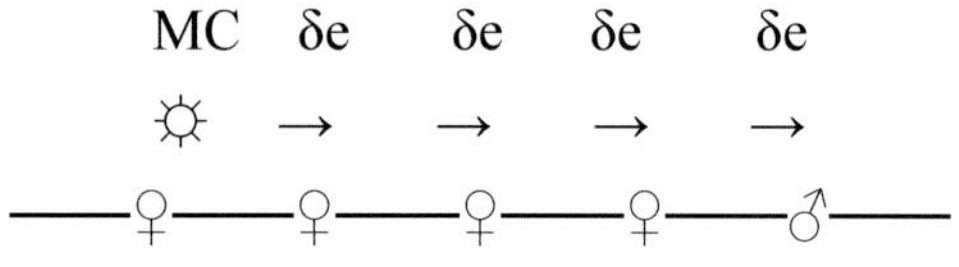

$$MC \quad \delta e \qquad \delta e \qquad \delta e \qquad \delta e$$

$$\text{Polarization growth, } P_{sum} = \Sigma p_{fg} + pMC$$

Designations: MC—mesocomposite; δe—charge quant (electron); → – the polarization direction; ♀——♀ – macromolecule fragment with functional groups; P_{sum}—summary polarization of system; Σp_{fg}—sum of polarizations of functional groups; and pMc—polarization (or dipole moment) of mesocomposite.

FIGURE 11.3   The scheme of polarization at charge quantization with the expansion of quants' influence on the polar groups of polymeric materials.

Thus, the mesocomposite action on media or compositions can be presented as the action of electron quants flow from mesoparticle to correspondent functional polar groups and chemical bonds of initiated substances.

Certainly, this flow cannot be stationary and will be changed at the mesoparticle quantitative change within macrosystem (or substance modified).

The phase coherency disturbance and the polarization effect for the mesoparticles (modifiers) in suspensions or polymeric compositions are possible at technological processing, for example, at ultrasound actions on suspensions. This operation is carried out for guarantee of equal distribution of mesoparticles in suspension.

For instance, the essential growth of intensities relation takes place at the ultrasound action on the copper-carbon mesocomposite suspension in isomethyl tetrahydrophthalate anhydride in the course of 7 min and also with the duration in 10 min. In these cases, the concentration of mesocomposite in suspension corresponds to 0.001% (Fig. 11.4 and Table 11.3).

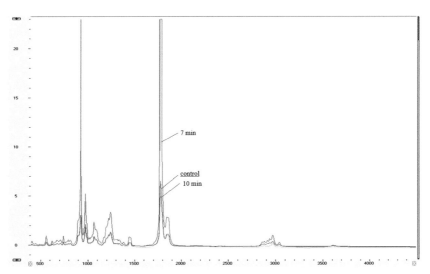

• time of ultrasound action: 7 min; • without ultrasound action; • time of ultrasound action: 10 min

FIGURE 11.4   The changes in IR spectra for fine dispersed suspension of copper-carbon mesocomposite (ω = 0.001%) on the basis of isomethyl tetrahydrophthalate anhydride depending on the duration of ultrasound processing.[8]

In Table 11.3, the comparison of changes in intensity relations at optimal time of ultrasound action on the suspension (7 min) and at the increased time (10 min) is adduced.

TABLE 11.3 Characteristic Wavenumbers and Peaks Intensity Relations at the Different Time of Ultrasound Action.[8]

N	ν (cm^{-1})	I_7/I_Θ	I_{10}/I_Θ	Conformity
1	1776.6	3.7932	0.7574	C=O st as
2	1844.1	2.5065	0.9115	C=O st sy
3	3039.1	2.3849	0.9589	C–H st

The increase in intensity relations for considered groups at the action duration equaled to 7 min is presented as follows: for C=O (unsymmetrical vibrations)—in 3.79-times; for C=O (symmetric vibration)—in 2.51-times; for C–H bonds (at 3039 cm^{-1})—in 2.385-times.

The great value of the intensity relation growth for C=O (unsymmetrical vibrations) can be explained by the increase in the dipole and the dipole moment of this bond. In other words, it is possible that the transformation of C=O bond on the equivalent on composition bond $C^{+\delta}{\rightarrow}O^{-\delta}$ companied with the polarity and dipole moment growth. Therefore, the probability of group interactions with their orientations concerning each other, hence the probability of self-organization is increased.

However, the ultrasound action on suspensions can initiate the disturbance of molecules orientation and leads to the self-organization absence. In our case, the increasing period of ultrasound action to 10 min leads to sharp downfall of peak intensities relations: for C=O (as)—in five-times; for C=O (sym)—in 2.75-times; for C–H bond (3039 cm^{-1})—in 2.5-times. This sharp downfall of peak intensities in IR spectra can be explained only by the phase coherency disturbance with the destruction of mesoscopic systems formed at the polarization under the mesoparticles influence.

Thus, the optimal result on IR spectra, for example presented suspension of "Cu-C MC – iMTHPhA" is arrived when the concentration of correspondent mesocomposite is equal to 10^{-3}%, and the ultrasound action duration corresponds to 7 min. If the ultrasound action is equal to 10 min, then the sharp downfall of the peak intensities in IR spectra is observed.

It is probable that this fact is explained by the phase coherency disturbance because of the energetic action growth with the increase in the ultrasound action duration.

11.3   ABOUT CHOICE OF LIQUID MEDIA FOR NANOSTRUCTURED MODIFIERS

The process of the quants flow interaction with the dipole of molecules in compositions (or media) on our mind, is analogous to inductomeric effect, when the length of dipole for polar group is increased, which can be expressed by following equation:

$$p_{\text{инд}} = p_0 + p_{\alpha} \tag{11.6}$$

where, $p_{\text{инд}}$—dipole moment, increased under quants flow action; p_0—the value of initial dipole moment; p_{α}—additional contribution in the value of dipole moment at the expense of group (fragment) polarization under the action of quants from the mesoparticle modifier.

The increase in dipole moments of the medium (or composition) functional groups leads to the interaction energy growth between dipoles, that is defined to inequality:

$$\varepsilon_{\text{ind}} = p_{1\text{ind}} \cdot p_{2\text{ind}}. / r_{\text{ind}}^{3} > \varepsilon = p_1 \cdot p_2 / r^3 \tag{11.7}$$

where, ε and p with index "ind" corresponds to the meanings of energy and dipole moments increased under the quants flow action; and r_{ind}—the distance between dipoles polarized.

It is possible that the interference appeared with new chemical bonds formation at the approach of dipoles of the functional groups of macromolecules.

Thus, the main result of materials' modification with the application of nanostructures can be the change in the density of materials. The density of materials is increased corresponds to the durability growth. This dependence between the named properties takes place, especially, at the modification of building materials by carbon nanotubes.[4–6]

Below the example of carbon nanotubes application for the modification of hard concrete and gas concrete is considered. Usually for the modification of concretes, the water suspensions of carbon nanotubes are used. The carbon nanotubes as good products are the produced mesoparticles (carbon nanotubes) covered by the carboxyl cellulose. These mesoparticles are introduced into the building compositions as the water suspensions.

The hardening of compositions is realized by the water vaporization during the certain quantities of days, for example, the hardening of gas concrete is produced during 56 days. It is necessary to note that at the beginning the water polarization takes place under the influence of nanostructures. Then at the water vaporization, the polarization effect is decreased that leads to the decrease in durability for the nanostructured material (concrete). When the drying of concrete composition is produced, the relation (R) of destructive tensions at compression for modified and nonmodified compositions is decreased (Figure 11.5, curve a).

In our case, the positive effect is decreased after 56 days of the drying from $R = 1.8$ to 1.1.

Thus, the improvement of durability characteristics by means of the introduction of carbon nanotubes in water suspension is realized only on 10%. Certainly, this result does not satisfy consumer.

Analogous effect of decrease in stability at the composition drying is observed for the hard concrete compositions (Figure 11.5, curve b).

The effect of decrease in physic mechanic indexes improvement can be explained by the phase coherency disturbance because of medium Zeta-potential change in composition.

The introduction of superficially active substance, for instance, lignosulphonate, leads to partial conservation of carbon nanotubes positive action on composition. Analogous effect is obtained if modifier mesoparticles are introduced into composition by means of its components fine dispersion.

The stable effect of material properties improvement is possible if the modifier mesoparticles are introduced in composition using the fine dispersed suspensions on the basis of liquid components such as plasticizer, curing agent, or binding.

Then, the positive effect is stipulated by increase in the self-organization effect during the formation of polymeric net structures.

The composition density growth is provided by the action of quants flow on the components. These quants refer from the mesoparticles to correspondent components of composition at the hardening which the polarization effect is "consolidated".

Then, the new denser phase or phase zone is organized. In the same cases, these zones are reinforced layers.[6, 7]

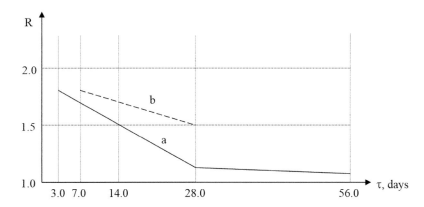

FIGURE 11.5   The relation change of durability for modified and nonmodified gas concrete (a) as well as hard concrete (b) depending on the sample drying duration. R—the relation of destructive tension at pressure for nanostructured material to correspondent parameter of material that did not modified by carbon nanotubes.

The new phase's formation leads to the growth of material's heat capacity. As the example of the influence of nanostructures on the density and thermal physical characteristics let us consider a sample of material that is obtained from cardboard saturated with the silicate glue that contains 0.003% nickel-carbon mesocomposite. The results of this sample studies are presented in Table 11.4.

TABLE 11.4   Thermophysical Characteristics of Modified Sample in Comparison with its Nonmodified Analog.

Properties	Samples	
	Nonmodified	Modified (0.003% MC)
Density, kg/m³	624.5	669 (↑7%)
Heat capacity, kJ/kg·K	1.79	2.972 (↑66%)
Heat conductivity, W/m·K	0.083	0.064 (↓23%)

The modifier mesoparticles are distributed in the suspension on the basis of silicate glue by means of ultrasound action.

The investigations of the sample thermal properties lead to the following results:

(1) The temperature conductivity is decreased on 43%.
(2) The density is increased on 17%.
(3) The heat capacity is increased on 66%.
(4) The heat conductivity is decreased on 23%.

The increase in adhesion strength is noted practically in all investigations dedicated to the modification of materials by nanostructures. These facts can be explained by the activation of quants flow from the mesoparticles on the phase board in system "adhesive–substrate". In this case, phenomena such as polarization and orientation of functional groups and active centers of adhesive and substrate take place. In these conditions, the interference is appeared and the chemical bonds are formed, which stimulates the formation of phase dense zone that makes sure the stable link between adhesive and substrate. However, this is possible when the polarization changes during modification process with use of nanostructures are absent.

Thus, for the obtaining of stable positive effect at the modification of materials by nanostructures or mesocomposites, it is necessary to exclude the phase coherency disturbance and to promote the appearance of interference. Also, the changes in the nature of nanostructures and also the relation in system "nanostructure–medium" must be absent.

11.4 CONCLUSIONS

In the present chapter, the possible reasons of the decrease or absence positive effects at the nanostructures or mesoparticles introduction into polymeric compositions, including concretes, are discussed.

On the basis of experimental data comparison, it has proposed the main reason of negative results at the application of nanostructures for the materials' modification as the phase coherency disturbance. This disturbance can be undertaken by following moments: (1) the change of modifier (nanostructure) in composition; (2) the technological procedure of ultrasound dispersion for the fine dispersed suspension producing; and (3) the vaporization of polar component that is the part of medium polarized by nanostructure in the suspension.

At present, the correspondent ways working out on the application of active nanostructures for the characteristics of materials' improvement becomes a significant trend of the mesoscopic industry development.

KEYWORDS

- **phase coherence**
- **mesoscopic modifiers**
- **metal carbon mesocomposite**
- **carbon nanotubes**
- **polarization**
- **modification**
- **z potentials**
- **IR spectra**
- **polymeric compositions**
- **concrete**

REFERENCES

1. Moskalets, M. V. Fundamentals of Mesoscopic Physics. Khar'kov: NTU KhPI. 2010, p.180.
2. Kodolov, V. I.; Trineeva, V. V.; Vasil'chenko, Yu. M. The Calculating Experiments for Metal/Carbon Nanocomposites Synthesis in Polymeric Matrices with the Application of Avrami Equations. In *Nanostructures, Nanomaterials and Nanotechnologies to Nanoindustry*; Apple Academic Press: Toronto, New Jersey, USA, 2015; 400, pp 105–118.
3. Trineeva, V.V. Technology of Metal Carbon Nanocomposites Obtaining and their Application for Modification of Polymeric Materials. Ph.D Thesis, sci. Kazan, 2015, p 41.
4. Yakovlev, G. I. Structural Organization of Interphase Layers at the Creation of Crystal Hydrate Composite Materials. Ph.D. Thesis, sci. Per'm, 2004, p 35.
5. Khokhriakov, N. V.; Yakovlev, G. I.; Kodolov, V. I. Modeling of Hydration of Calcium Sulphate Hemihydrate. *Chem. Phys. Mesoscopics* **2000**, *2*(2), 205–213.
6. Korablev, G.A.; Yakovlev, G. I.; Kodolov, V. I. Some Peculiarities of Cluster Formation in System $CaSO_4 – H_2O$. *Chem. Phys. Mesoscopics* **2002**, *4*(2), 188–199.
7. Kodolov, V. I.; Lipanov, A. M.; Trineeva, V. V. et al. The Change of Properties of Materials Modified by Metal-Carbon Nanocomposites. In *Nanostructure,*

Nanosystems and Nanostructured Materials: Theory, Production and Development; Apple Academic Press: Toronto, New Jersey, USA, 2014; 558, pp 327–373.

8. Kodolov, V. I.; Kodolova–Chukhontzeva, V. V. Fundamentals of Chemical Mesoscopics. Izhevsk: Publ. M.T. Kalashnikov IzhSTU, 2019, 220 p.

Index